Sherrington's Loom

Sherrington's Loom

An Introduction to the Science of Consciousness

ALAN J. McCOMAS

(with drawings by Marie Levesque)
Neuroscientist and Emeritus Professor of Medicine,
McMaster University, Hamilton, Canada

Oxford University Press is a department of the University of Oxford. It furthers
the University's objective of excellence in research, scholarship, and education
by publishing worldwide. Oxford is a registered trade mark of Oxford University
Press in the UK and certain other countries.

Published in the United States of America by Oxford University Press
198 Madison Avenue, New York, NY 10016, United States of America.

© Oxford University Press 2020

All rights reserved. No part of this publication may be reproduced, stored in
a retrieval system, or transmitted, in any form or by any means, without the
prior permission in writing of Oxford University Press, or as expressly permitted
by law, by license, or under terms agreed with the appropriate reproduction
rights organization. Inquiries concerning reproduction outside the scope of the
above should be sent to the Rights Department, Oxford University Press, at the
address above.

You must not circulate this work in any other form
and you must impose this same condition on any acquirer.

CIP data is on file at the Library of Congress
ISBN 978-0-19-093654-9

To my lovely wife, Marie Ambruz

Contents

Preface   ix
Vignette   xiii

1. Plan of the Book   1
2. Definition and the Mind–Body Problem   4
3. Concepts, Conferences, and Controversies   12
4. Who Else Is Conscious?   32
5. Groundwork: Cellular Events   49
6. Mapping the Brain   69
7. Awaking the Cortex   103
8. Electricity Works   123
9. Single Units and Grandmother Cells   141
10. Making Sense of the Senses   161
11. Benjamin Libet's Big Experiment   186
12. More on Gnostic Units and Cortical Columns   200
13. Continuing the Synthesis   226
14. Looking to the Future   254
15. Summary   264
16. Postscript   267

Acknowledgments   269
Index   271

Preface

For me, the fascination of neurophysiology began with the first lecture to the medical student class. It was a new world, one of electric impulses, and its heroes were Hodgkin and Huxley, who used an electrode inside the giant axon of the squid as their weapon. There was also mention of an older man, the seemingly omniscient Sherrington with his concepts of central excitatory and inhibitory states in the spinal cord—and mention, too, of subsequent microelectrode recordings in single motoneurons by Sherrington's very able former student, Eccles (a laugh here among those students who listened to *The Goon Show* on the radio). Every lecture added to the intellectual excitement, as did the practical classes when the twitches of the frog's legs were enthusiastically recorded on smoked drums.

Later there came a time to study neurophysiology more seriously and to observe and meet some of its most able practitioners at meetings of the Physiological Society in the United Kingdom. For aspiring neurophysiologists like myself, John Fulton's *Physiology of the Nervous System* and Bernard Katz's slender paperback, *Nerve, Muscle and Synapse*, became our Old and New Testaments, respectively. During this period I was indebted to Professor John Gray at University College London for encouraging me in my first real research (this on the gracile nucleus of the rat). Later still, in Newcastle upon Tyne, there was the opportunity to establish a muscle laboratory as part of the muscular dystrophy research program headed by Dr. John Walton (subsequently Lord Walton of Detchant). There was one day in the week, however, when muscle was set aside in favor of recordings from the thalamus during stereotactic surgery on patients suffering from parkinsonism or intractable pain. This was in the late 1960s, and soon the surgery and the recordings would be things of the past as treatment with L-dopa took over. But the thalamic recordings had left their mark, not least of which was the nagging puzzle as to why it was impossible to influence the firing of the great majority of cells encountered by the exploring electrode!

Even now, many years on, the fascination is still there. How could it not be? The voltage clamping methodology that yielded the data for the Hodgkin–Huxley equations, a methodology that was so remarkable in its day, has been followed by a series of astonishing techniques. Insertion of voltage-sensitive dyes now enables nerve fibers to glow when conducting impulses. And light of various wavelengths can stimulate neurons that have been genetically modified to contain photosensitive pigments. On a larger scale, active areas in human brains are now immediately identifiable by functional magnetic resonance imaging or magnetoencephalography, while microelectrode recordings in the brains of conscious human subjects can be made to drive robotic limbs, or to locate memories of people, places, and events. And for neuroscientists disinclined to work in the laboratory there is the option of mathematical modeling as a means to explore how neural circuits might operate.

With such a rich variety of techniques for the neuroscientist to draw upon, it was inevitable that consciousness should continue as a prime target for investigation. Long the property of philosophers, the subject was never far from the minds of neuroscientists even though the first major symposium on the topic did not occur until 1953. Most likely, it was the discovery of the reticular activating system by Magoun and Moruzzi soon after the end of World War II that had been the main catalyst. Since that time the publications on consciousness have increased exponentially, barely a year passing without an international symposium on the topic. In addition, specialist journals largely devoted to consciousness have multiplied, as have the number of books. Somewhere, buried in this ever-rising mountain of data, much of it highly technical, are the clues for understanding the neural activity responsible for consciousness. For a beginner especially, the challenges of finding these clues and of appreciating their significance are daunting. It is for just such a person that this book has been written.

I have used a historical approach, partly because it marks the paths taken but also as a means of honoring those whose work enabled the present positions to be reached. One especially admired was Herbert Jasper, who graciously introduced himself across a breakfast table in Kingston, Ontario, while attending a retirement function at Queens' University. It was Dr. Jasper who responded to my inquiries as to Penfield's reaction to the attack made on his work by Francis Walshe, and it was he who kindly invited me to participate in the 1997 Montreal symposium on consciousness.

I wish I could have met Sherrington, whose work has been used to open and close the book. That giant, formerly recognized as the greatest physiologist of his time, died at a great age just as I was starting medical school. A diligent pilgrim, I located his retirement house in Ipswich, read his laboratory notes in Vancouver, and worked at his former desk in Oxford. One great man whom I did meet, without being aware of his importance at the time, was Jerzy Konorski—this during a visit to Warsaw in the very cold winter of 1968–1969. It is Konorski's well-reasoned ideas concerning the presence and activities of "gnostic" units that, together with cortical columns and time-chunks, are the real core of the later section of this book. I should add that the time-chunk phenomenon and its implications for the functioning of cortical columns became an interest following some experiments on backward masking.[1,2] Of my other researches, I would only mention an exploration of Quebec's Laurentian mountains in order to find the Alpine Inn, the site of the 1953 international symposium on Brain Mechanisms and Consciousness. Looking at the inn, no longer the premier establishment it once was, and reflecting on the stature of those who had assembled there that August, it was difficult to understand later claims to have "invented the science of consciousness."[3]

The use of a historical approach in tackling consciousness brings with it a difficulty altogether absent in writing *Galvani's Spark*. That book, describing the development of research on the nerve impulse, has a linear narrative. The impulse research was stretched over two centuries and involved relatively few laboratories, only one of which was dominant at any given time. Moreover, the description of research into the impulse has a definite starting point, with the accidental twitch of a frog's leg in a house in Bologna, Italy. Just as important, the story has a logical conclusion, with Roderick Mackinnon's elucidation of the three-dimensional molecular structure of the ion channels in nerve membranes. In contrast, the scientific pursuit of consciousness has no obvious beginning and the end has yet to come, if ever it does. And instead of a single laboratory dominating, the consciousness research field is worldwide and continues to expand—as does the number of publications.

Rather than covering everything to do with consciousness—an impossible task—I have concentrated on a small number of observations of major significance. Even though the approach is a simple one, there are bound to be some factual errors, and it would be surprising if there were not disagreements over content and interpretations of data. Nor is there any claim to originality, for I suspect that some or all of the ideas in the final chapters have been entertained by others at one time or another. Regarding the illustrations, ideally there would have been several additional portraits of neuroscientists, for surely those of Roger Sperry, Ralph Gerard, Victor Hamburger, Rita Levi-Montalcini, and Donald Griffin deserved their places. Alas, either the copyright ownership was not known or the fees for reproduction were prohibitive. In this context one can only regret the insertion of a third party into the permissions process and, with it, the imposition of substantial fees for material that had previously been freely, and often graciously, shared with inquiring authors. As will become obvious from figure captions, I owe a huge debt to Wikimedia Commons and the Wellcome Collection, and it is to be hoped that the continued expansion of Wikimedia Commons will largely obviate the need for third-party fees.

Lastly, one cannot finish a book of this kind without expressing a conviction shared by most, if not all, of those studying the brain. Although much more is now known about the way the brain functions, there should not be any lessening of admiration for this remarkable organ—this physicochemical machine that is not only capable of investigating itself but of examining everything else in the universe as well. How could there be anything more wonderful and, for a neuroscientist, anything more beguiling, than these 1.3 kilograms of gray matter!

Notes and References

1. MacIntyre NJ, McComas AJ. Non-conscious choice in cutaneous backward masking. *NeuroReport*. 1996; 7: 1513–1516.
2. McComas AJ, Cupido CM. The RULER model: Is this how the somatosensory cortex works? *Clinical Neurophysiology*. 1999; 110: 1987–1994. Note: the time-chunk phenomenon was attributed to the "reading out" of columnar activity but was mistaken in proposing feedback via axon collaterals.
3. For example: Koch C. The footprints of consciousness: the bits and pieces of the brain reside in places few suspected. *Scientific American Mind*. 2017; 28: 52–59.

Vignette

Ipswich, England
November 1936

The small man put down his pen, sat back in his chair, and took off his spectacles. He had been writing for an hour and now he needed to rest his mind. The desk was much smaller than the one he had been used to in Oxford, that great double-sided desk with its many drawers that had served him so well over the years. He thought back to those days, remembering the young men who had come to work with him in the laboratory so as to learn about the brain. Eric Liddell. Derek Denny-Brown from New Zealand. William Gibson from Canada. John Eccles from Melbourne. Yes, Eccles especially. What energy that young man had—and how thoughtful of Eccles to have visited him earlier in the week. A kind heart hidden behind a brash exterior. And as argumentative as ever when they had discussed their old work!

He wondered what Eccles had thought about the new home of his former mentor. A modest little house on a very undistinguished street, so different from Oxford. But Ipswich was where he had grown up, and so it was fitting that he should have chosen the Suffolk town for his final years. He remembered his schooldays and when, under the kind influence of his stepfather, he had started to think about the brain.

The brain. How could he find the words he needed? The words that would help him through the lectures he was soon to give in Edinburgh. How could he convey the wonder and magnificence of the brain to a new audience?

Sherrington glanced up. It was growing dark outside and through the window he could see lights appearing in the houses beyond his garden. Lights, yes, perhaps . . . He picked up the pen and began to write. Now another image formed in his mind. Not Ipswich, not Oxford, but the Lancashire woollen mill he had visited during his days in Liverpool. All those women, young and old, standing as they worked the looms. Yes, those looms with the shuttles flying from side to side and the pattern in the cloth appearing as if by magic. As he thought back, so the words came to him.

> The great topmost sheet of the mass, that where hardly a light had twinkled or moved, becomes now a sparkling field of rhythmic points with trains of travelling sparks hurrying hither and thither. The brain is waking and with it the mind is returning. It is as if the Milky Way entered upon some cosmic dance. Swiftly the head-mass becomes an enchanted loom where millions of flashing

shuttles weave a dissolving pattern, always a meaningful pattern, though never an abiding one.

Sherrington had a facility with words. The gift had revealed itself in his scientific papers as well as in his poems. In his long life he had been knighted by his king, had been the recipient of a Nobel Prize, and, in his prime, was recognized throughout the scientific world as the greatest authority on the nervous system.

Now in the twilight of his life, there was still much to say. Gradually the lectures took their shape—the lectures that would be his last major work.[1]

Charles Scott Sherrington (1857–1952)

Starting the book with a vignette is a means of honoring the person who, for much of his long working life, dominated thought, discussion, and experiment to do with the nervous system. Born in London but growing up in East Anglia, Sherrington was heavily influenced by his stepfather, a physician with a strong interest in the arts. After his schooling, Sherrington began medical studies at St. Thomas' Hospital in London but then transferred to Cambridge, where he was to gain first-class honors in both parts of the Natural Tripos. While still a student, he attended the International Medical Congress in London and was especially interested in David Ferrier's evidence for the presence of a motor area in the monkey brain. Next came a period of study in Germany with the neuropathologist, Rudolf Virchow, followed by a return to London and a lectureship at St. Thomas' Hospital Medical School. In 1885 Sherrington was appointed professor of physiology at University College, Liverpool, where Richard Caton had been the first chair. It was in Liverpool that Sherrington did much of his best work, including the study with Albert Leyton in which the details of the cortical motor representation were mapped out in several primate species (see main text). By now famous, Sherrington was sought after, and visited by, clinicians and scientists working on the nervous system. His reputation was consolidated by an invitation to become the Waynflete Professor of Physiology at Oxford, a post he assumed in 1914. From that time onward, Sherrington concentrated his laboratory work on the exploration of the spinal cord, using the spinal reflexes to investigate central excitation and inhibition. During this period, which continued until his retirement in his 70s, Sherrington had a number of graduate students who subsequently achieved distinction themselves, two of them (John Eccles and Ragnar Granit) winning Nobel Prizes; Sherrington himself had shared the Prize with Edgar Adrian in 1932. Like a number of other neurophysiological pioneers, Sherrington had a long life, dying in a retirement home at 94; right until the end, however, this small man, the recipient of numerous awards and honorary

Figure V.1. Charles Scott Sherrington, possibly taken during his Liverpool years. (Wellcome Collection)

degrees and the greatest living authority on the nervous system, remained in full possession of his faculties and was still able to discuss the workings of the brain with his visitors.

Note and Reference

1. Sherrington CS. *Man on his nature*. The Gifford Lectures, Edinburgh 1937–1938. Cambridge, England: Cambridge University Press, 1940.

1
Plan of the Book

As already stated, the present book is an introduction to the science of consciousness and is intended for newcomers to the field as well as those interested in the development of neuroscience. Though essentially historical, the book contains several novel concepts.

The next chapter is brief and is concerned with definition and the monism-dualism debate. This is followed by the history of research meetings dealing with consciousness, beginning with that hosted by Herbert Jasper in the Laurentian mountains of Quebec in 1953. It also touches on a central debate that reached its climax a little later, as to which part of the brain was responsible for consciousness. Was it the cerebral cortex, as had been the prevailing assumption, or was it the brain stem? As will become obvious, there was already much consciousness research going on before Francis Crick's entry into the field in 1979.

In the next chapter it is the studies of the behaviors of various animal species that are important, for these indicate that they, the animals, share the property of consciousness with us humans. It is something that we should have known and thought about in shaping our attitudes to animals, especially towards those used in medical research laboratories. But the chapter does something else—it draws awareness to the fact that brains do not have to be built like ours in order to possess consciousness.

The succeeding chapter helps to prepare the ground for much of what is to follow. It briefly traces the history of research into the structure of the nervous system and into the nature of the nerve impulse, the transitory event that enables neurons to communicate and upon which the brain—Sherrington's "enchanted loom"—depends for all its transactions. There is mention, too, of the synaptic connections that determine the excitatory or inhibitory responses of the nerve cells to incoming impulses.

After the elements comes a consideration of the whole. How was it possible to deduce the functions of the different parts of the brain, and what did the classical brain maps look like? As will become clear later in the book, the major excitement now is no longer in the sensory receiving areas and motor cortex but in what had been loosely termed the "association" areas, those large tracts of cortex whose functions were long shrouded in mystery.

Next comes the story of the discovery of the reticular activating system, a discovery that provided an enormous impetus to brain research, and especially into conscious mechanisms. But, even before Moruzzi and Magoun's seminal paper, there was evidence that the general activity of the cortex was heavily dependent on that from below, and from the non-specific thalamic nuclei in particular.

The electrical activity of the brain is the subject of the following chapter, the starting point being Richard Caton's demonstration of slow waves in the rabbit brain before an audience of physicians in Edinburgh in 1875. Then comes mention of the impact of Berger's discovery of similar slow waves in the human brain and of the advent of EEG (electroencephalography). The chapter finishes with the remarkable technical accomplishment of Mircea Steriade in being able to record from the same single neuron during periods of sleep and wakefulness, thereby showing the enormous range of impulse firing frequencies possible. Is it possible that it is simply the intensity of the cortical discharge, with its thalamic underpinning, that determines whether or not impulse activity enters into consciousness?

Next comes a truly important recent discovery in the human hippocampus, that of "concept" (or "grandmother") cells—neurons that code for multiple aspects of the same person or object. The prediction that specific recognition cells were present in the brain had been made many years previously by vision scientists in MIT and Cambridge University. At about the same time Jerzy Konorski, in Warsaw, had argued for the existence of similar neurons ("gnostic units") serving a number of functions.

The succeeding chapter follows logically, being an excursion into our most important sense, that of vision. It describes the novel findings of David Hubel and Torsten Wiesel when recording from single cells in the primary visual cortex, and how these findings supported the concept that the various features of the observed image underwent independent processing in parallel. However, the chapter also presents an alternative to the now classic Hubel-Wiesel scheme, one that, despite its fundamental differences, seems equally plausible.

The following chapter describes the line of research that led to one of the most important experiments in consciousness research—Benjamin Libet's finding that electrical activity in the brain precedes conscious awareness. Though it runs counter to our common sense and ingrained belief, free will appears to be an illusion!

The book concludes with three chapters, one of which draws attention to another key observation, the remarkable differences between the left and right hemispheres, as judged by their contrasting behaviors in split-brain patients. The main purpose of these last chapters, however, is to attempt to incorporate the old and new findings into a scheme for the neural events underlying consciousness. Regarding some of the novel features, it would be well to recall the Royal Society's injunction at the time of its sponsorship of Robert Hooke's *Micrographia* in 1664—"the several hypotheses and theories laid down . . . are not delivered as certainties, but as conjectures."[1] The term "conjecture" is an appealing one as it avoids the formalism of "hypothesis." Instead, its Latin roots stress the "putting together" of evidence even though, in the case of the brain and its relation to consciousness, the evidence is mostly circumstantial and therefore incomplete.

To a large extent the proposed scheme is based on the ideas of Jerzy Konorski—ideas that have been overlooked by those unfamiliar with his 1967 book, *The Integrative Activity of the Brain*. The reasons for adopting Konorski's concepts depend not only on his authority as an neurologist, neurophysiologist and psychologist, but also on the subsequent confirmation of his prediction of "gnostic" units present in the brain. It is surely significant that ideas similar to Konorski's have been developed independently by artificial intelligence pioneers such as Geoffrey Hinton and Ray Kurzweil,[2] and are now being employed successfully in "deep learning."

The final chapter, titled "Looking to the Future," was necessitated by the possibility of machines developing intelligent behavior and even achieving consciousness. So remarkable have been the developments in computer-based technology, that the issue cannot be ignored—even though it may prove incapable of being satisfactorily resolved.

By its nature, the book is as much about people as ideas and achievements. Consequently, where there has been insufficient material in the text about a particular neuroscientist, additional information has been provided in the form of "boxes." The choice of whom to include in a box has been somewhat arbitrary and there are Nobel laureates and potential laureates who are barely mentioned, but the emphasis has been on those no longer with us.

Notes and References

1. Gribbin J, Gribbin M. *Out of the shadow of a giant: Hooke, Halley & the birth of science.* New Haven, CT: Yale University Press, 2017.
2. Kurzweil R. *How to create a mind. The secret of human thought revealed.* New York, NY: Viking Penguin, 2012.

2
Definition and the Mind–Body Problem

Definition

Before the key observations can be considered, it is necessary to deal with two major issues—a definition of consciousness and the mind–brain problem. Of the two, the definition of consciousness is the easier task, but it is still far from straightforward, as reflected in the fact that at least 29 definitions have been proposed.[1] Rather than consider all of these, each of which has its own justification, we will adopt a simple but nevertheless profound definition; namely, *consciousness is an organism's awareness of itself.*

Note that use of the term *organism* makes room for the possibility that non-human species may also be aware of themselves—a possibility that will be considered in Chapter 4. The definition describes what might be considered as the most basic form of consciousness and is the one that, in the case of humans, is possessed by a newly born infant. There comes a time, however, when this basic consciousness is complemented by an ability to reason and plan. This sudden acquisition may be dramatic for the child, as in my own case:

> As a two-year-old toddler I had wandered away from the West Australian farmhouse where I lived with my parents. Spying my mother creeping behind a hedge toward me, I realized that, to avoid a scolding, my best strategy was to pretend I had not seen her and to sit down on the grass, calling out for her. At the time, I had a sense of wonder that I had been able to think of this course of action. I think it was the first occasion that I had heard an internal voice.

The ability to think ahead, without the accompaniment of an internal voice, probably comes much earlier, however.

> While taking my 12-month old daughter for daily walks in her pram, I became the butt of a game that she had devised for her own amusement. Thus, while I was deliberately looking elsewhere, she, sitting upright, would remove first one, and then the other, of the woolen mitts from her hands and drop them

over the side of the pram. My role was to make a great fuss on discovering the absence of the mitts, prior to retrieving them from the sidewalk. During this pantomime, she—the one-year old—would be chuckling with laughter. And the process would be repeated to similar delight a little later.

In later life, the awareness of the power of one's own thinking may lead to what may be considered the highest forms of consciousness, such as the ability to solve advanced mathematical and scientific problems, to think in abstract terms, and to create music and works of art.

The Mind–Body Problem

Not surprisingly, the relationship between the mind and the body is one that has exercised philosophers from earliest times, not least because it touches on the nature of reality. For the French mathematician and philosopher René Descartes (Box 2.1), the senses could not be trusted to provide objective information and the only reality that he could be sure of was his own existence: "*Cogito, ergo sum*" (I think; therefore, I am). Descartes visualized the mind ("soul") as independent of the body, though capable of influencing it. His choice of the pineal gland as the site of this interaction was largely because of its position in the midline of the head, close to the lateral ventricles. He envisaged the pineal controlling the flow of "spirits" (minute corpuscles) from the ventricles through openings in the surrounding brain matter and thence along the nerves to the various parts of the body (Figure 2.1).[2]

Although not stated, it is possible that the choice of the pineal for the action of the soul was influenced by the common feeling that the "I," the person looking out on the world, is situated behind the eyes. In the welter of neuroscientific observations that have accumulated in the past century especially, it is easy to be scornful of the selection of the pineal gland as an intermediary for consciousness, but it must be remembered that Descartes was writing four centuries ago and, among other accomplishments, had conceived an entirely new branch of mathematics. Further, our senses may indeed feed us false impressions of the outside world. Not only that, but many have followed Descartes in attempting to identify a particular region of the brain as the seat of consciousness. Most important of all, it was Descartes who stated the concept of dualism most clearly, that is, that the mind had a separate existence, being able to function independently of the body (i.e., brain). This, of course, is another way of saying that we have free will, and, as such, it is a belief that the great majority of the non-neuroscientific world, if asked, would probably subscribe to today.

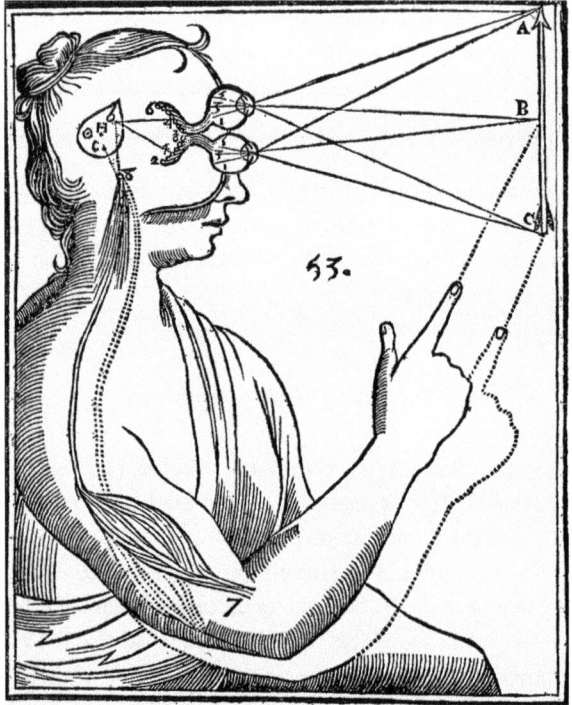

Figure 2.1. Descartes: coordination of muscle and visual mechanisms (from *L'homme et un traitte...*). In this figure Descartes has cleverly shown the interaction of the visual and motor systems. Initially the subject's gaze is directed downwards to a point, C, indicated with the faint outline of the index finger. By then contracting the biceps muscle, via nerves running to the arm from the pineal gland, he is able to flex the elbow and indicate a new point, B, directly ahead (seen with the image at a different location at the back of the eye). (Wellcome Collection)

The debate about dualism versus monism (that there is only one unifying reality) is one that continues even now and that the present book will not pursue directly. Instead we will point to another early thinker who, like Descartes, had a well-defined concept that he was able to express clearly—Thomas Huxley (Box 2.2). Huxley was a great admirer of Descartes because of the latter's ability to conceive of the body and brain as parts of a machine and thereby subject to laws governing the operations of machines; even the senses and the production of movements could be considered in this way—though not the "soul," of course. Though Huxley did not voice his disagreement with Descartes over the existence of a separate soul, he made it clear that he regarded consciousness as an epiphenomenon. Thus a mental event was a product of the brain's activity but was

Box 2.1 René Descartes (1596–1650)

René Descartes was born on March 31, 1596, in the small French town of La Haye. His family appears to have been moderately wealthy, although, because of his mother's early death, René was brought up by his grandparents. Having attended a Jesuit school, Descartes then entered the University of Poitiers and studied law. After a period of military service, Descartes spent the next years (1619–1628) either in Paris or traveling around Europe, and it was at this time that he began his philosophical studies. His overriding aim appears to have

Figure B2.1. René Descartes. Line engraving by C. Ammon, 1654. (Wellcome Collection)

> been to explain as much of the world, including the humans inhabiting it, in terms of scientific principles, and he pursued this ambition more thoroughly in the Netherlands during the next phase of his life. To improve his knowledge of anatomy, Descartes would frequently visit local slaughter houses, selecting body parts that he would bring home for careful dissection. Although he began writing his treatises during his sojourn in the Netherlands, he hesitated to have the works published for fear of incurring the displeasure of the Catholic Church—he was apparently aware of Galileo's difficulties, and it was only after his own death that Descartes's work appeared in print. As well as being a philosopher, Descartes was a formidable mathematician, his greatest achievement being the demonstration that any algebraic equation could be represented graphically, with the various terms having coordinates on vertical and horizontal axes; it was a development that foreshadowed the development of calculus. The end of Descartes's life came during a visit to Sweden, where he had been invited by Queen Christina to teach her philosophy and to set up a scientific institute. Having attended the French ambassador, a friend, who was suffering from pneumonia, Descartes developed the same condition himself and died on February 11, 1650.

unable to have any effect on the latter. In a passage from a famous essay, he drew an analogy with a steam locomotive:

> The consciousness of brutes would appear to be related to the mechanism of their body simply as a collateral product of its working, and to be as completely without any power of modifying that working as the steam-whistle which accompanies the work of a locomotive engine is without influence upon its machinery.[3]

Consciousness, then, was an epiphenomenon of brain activity. Needless to say, Huxley's vigorous arguments, buttressed by observations on the considerable behavioral abilities of animals and on the effects of brain damage in humans, aroused strong reactions in the scientific and philosophical communities. However, far from dualism being quenched, the latter continued to have strong advocates, one of them the 1963 Nobel Laureate, John Eccles. Eccles, a neurophysiologist, conceived of consciousness acting on the brain as a mental force incapable of measurement by existing physical instruments.[4] He suggested that synaptic knobs might be the target for such a force, pointing out that any effect would be greatly multiplied because of the extensive connections made by even a single neuron. Returning to the subject much later, and drawing on advances in understanding the fine anatomy of the cortex, Eccles introduced the concepts

Box 2.2 Thomas Henry Huxley (1825–1895)

TH Huxley was born in London to parents of modest means, his father being a schoolteacher. After attending his father's school, Huxley studied medicine at Charing Cross Hospital and then, after passing his initial exams, joined the Royal Navy. After months of inactivity, Huxley was appointed ship's surgeon on *HMS Rattlesnake* for a voyage to explore and chart the waters separating northern Australia from New Guinea. During this four and a half year expedition, he had the opportunity to study various life forms with the aid of a microscope and to write up his observations in the hope they would reach the Linnean Society. On returning to Britain, however, he had little hope of advancement and left the Navy, giving occasional lectures and preparing a full account of his scientific observations. And then, unexpectedly, he was asked by *The Times* to review a new book—the *Origin of Species* by a certain Charles Darwin. Huxley, a self-trained physiologist and a brilliant comparative anatomist, at once recognized the importance of Darwin's conclusions and, from

Figure B2.2. T. H. Huxley. Photograph by Elliott & Fry, London. (Wellcome Collection)

that moment on, became the leading spokesman for evolution and for the inclusion of Man in that process—needless to say, antagonizing the Church for his heretical views. Among his own scientific achievements was the deduction that birds had evolved from reptiles by way of small dinosaurs and his realization of the importance of the fossil record in tracing this and other evolutionary paths.* As Huxley came to be recognized as his country's foremost spokesman for science on the basis of his numerous lectures and publications, there followed appointments as Professor of Natural History at the Royal School of Mines and as Professor of Comparative Anatomy and Physiology at the Royal College of Surgeons. With his great intelligence, and his energy and skill as speaker and writer, Huxley later became president of both the Royal Society and the Royal Geographical Society, as well as a member of Queen Victoria's Privy Council. His international standing as a champion of science resulted in a visit to America. Among Huxley's gifted descendants were his grandsons Aldous (author of *Brave New World*), Julian (first director-general of UNESCO), and Andrew (Nobel Prize with Alan Hodgkin in 1963, for nerve impulse studies).

*The deduction was far from straightforward. See Switek B. Thomas Huxley and the dinobirds. Smithsonian.com. December 7, 2010. https://www.smithsonianmag.com/science-nature/thomas-henry-huxley-and-the-dinobirds-88519294/.

Source. Adapted from: Clark RW. *The Huxleys*. London, England: Heinemann, 1968.

of "psychons." The psychons were considered to be mental events, each of which had the capability of activating a corresponding cluster of apical dendrites in the cerebral cortex (a "dendron"); quantum mechanics accounted for the interaction between the two.[5]

The arguments for and against epiphenomenalism, as applied to consciousness by Huxley, have been admirably summarized elsewhere.[6] The position taken in this book is that Huxley's view was correct in the sense that the mind is a product of the working of the brain. Expressed simply, the brain, like biological structures in general, functions by means of physicochemical processes and is subject to all the laws governing the interaction of matter and energy. There is no special case for "mind." In Cartesian terms, the brain, though vastly complex, is as much a machine as is a muscle.

Where one might disagree, and do so vigorously, with Huxley is over the second part of his statement, namely, that consciousness is incapable of modifying the working of the body or, more specifically, the working of the brain. Thus, pain and pleasure are both products of consciousness and, by influencing the response of the body to evocative stimuli, shape future behavior in a manner beneficial to the organism and to the survival of the species. Reflexes, in the absence of consciousness, are insufficient. To take Huxley's analogy further, while the steam whistle contributes nothing to the running of the locomotive in the engineering works where it was built and tested, the warning of the whistle has positive value for the survival of the locomotive once it starts its duty on the rail tracks (and their level crossings).

Notes and References

1. Barŭss I. Metanalysis of definitions of consciousness. *Imagination, Cognition and Personality.* 1987; 6: 321–329.
2. Descartes R. *Traité de l'homme.* Paris, France, 1664. Translated by TS Hall. New York, NY: Prometheus Books, 2003.
3. Huxley TH. On the hypothesis that animals are automata, and its history. *The Fortnightly Review.* 1874; 15(ns): 555–580. Reprinted in *Method and results: Essays by Thomas H. Huxley.* New York, NY: Appleton, 1899.
4. Eccles JC. Hypotheses relating to the brain-mind problem. *Nature.* 1951; 168: 53–57.
5. Eccles JC. A unitary hypothesis of mind-brain interaction in cerebral cortex. *Proceedings of the Royal Society B.* 1990; 240: 433–451.
6. Robinson W. Epiphenomenalism. In: Zalta EN, ed. *The Stanford Encyclopedia of Philosophy.* Fall 2015 ed. https://plato.stanford.edu/entries/epiphenomenalism/

3

Concepts, Conferences, and Controversies

Hippocrates, Descartes, and Willis

A good starting point for considering the relationship between brain and mind would be the year 460 BCE, for it was then that the Greek physician, Hippocrates, was born on the island of Kos. Hippocrates was by no means the first physician of whom there is written record; that honor belongs to Amenhotep. Practicing in ancient Egypt two millennia before Hippocrates, Amenhotep was so revered as to have deserved a royal burial. Nevertheless it was Hippocrates who advanced medicine by freeing it from superstition and by making detailed studies of the patients under his care. In relation to consciousness the following clear statement appears among his writings:

> Men ought to know that from the brain, and from the brain only, arise our pleasures, joys, laughter and jest, as well as our sorrows, pains, griefs and tears. Through it, in particular, we think, see, hear, and distinguish the ugly from the beautiful, the bad from the good, the pleasant from the unpleasant.[1]

One could then jump two millennia to two major works that appeared close together. One was the posthumous publication, in 1662, of Descartes' *Treatise on Man*, already considered (see Chapter 1). The other, from the opposite side of the English Channel, was Thomas Willis' *Cerebri Anatome* (1664). Willis (1621–1675), while practicing medicine in Oxford—one of his patients had been King Charles I—had also undertaken careful dissections of brains that had belonged to criminals. Among Willis' later friends and colleagues was Christopher Wren, then Professor of Astronomy at Oxford, and it was Wren who made the magnificent etching of the undersurface of the human brain shown in Figure 3.1.

Impressed by the well-developed cerebral hemispheres in humans, Willis commented: "The cerebrum is the primary seat of the rational soul in man, and of the sensitive soul in animals. It is the source of movements and ideas[2]

Anatomy could only hint at function, however. A new era of investigation began with Luigi Galvani's discovery of "animal electricity" a century after Descartes' and Willis' publications. Galvani's experimental observations were the first in a series that, over the next 200 years, would ultimately show how electric

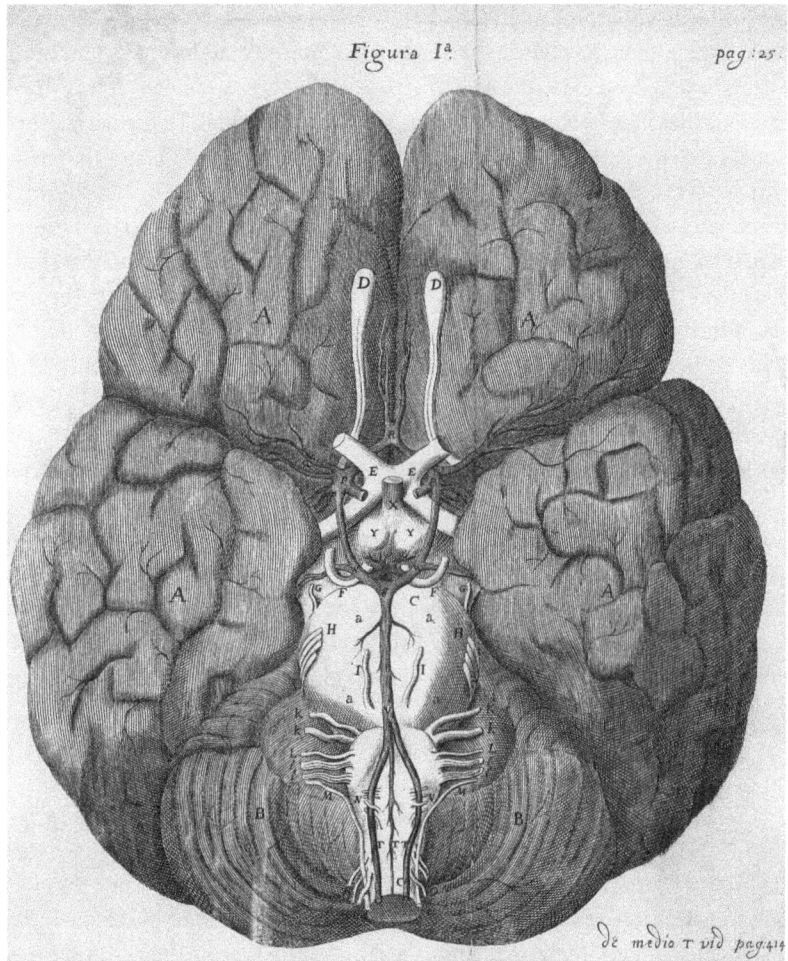

Figure 3.1. Christopher Wren's superb etching of the bottom of the human brain that appeared in Thomas Willis' *Cerebri Anatome*, 1664. (Wellcome Collection. CC BY 4.0)

impulses in neurons resulted from the brief opening and closing of ion channels in cell membranes. Among those who participated in the pursuit of the nervous system's electrical activity is an almost forgotten name, that of Richard Caton.

Caton's Discovery

At the time of his major contribution, Caton was a physician at the Royal Liverpool Infirmary in England. This was in the 1870s, a time when physiology was still in its infancy in Britain; moreover, Liverpool, being the country's main port for trade with the United States and Canada, was a city given over to commerce rather than to science. Nevertheless Caton, in addition to tending his patients, was a keen experimenter and possessed a galvanometer. It was this instrument that he had thought to apply, through electrodes, to the surface of an animal's exposed brain. It must have been with surprise and delight that he was able to observe the galvanometer needle moving, not only moving but altering its movements when the animal was stimulated; when the animal had died, the galvanometer needle was still.[3] Here, then, was the first demonstration that, just as Galvani and others had shown in the nerve fibers of the arms and legs, the nerve cells in the brain worked by electricity.

Since the brain had been identified by Hippocrates and, subsequently, by the likes of Descartes and Willis as the seat of the mind ("soul"), and since Caton had now shown that the brain appeared to operate electrically, it was a reasonable step to postulate that consciousness must, like the other functions performed by the brain, be the result of electrical changes produced by the brain's cells. In outline, then, the neural mechanism underlying consciousness had been found. It remained to determine which contributions the various parts of the brain made to consciousness and to discover how the electrical activity critical for consciousness was generated. It is these two basic questions that have preoccupied neuroscientists interested in consciousness during the 140 or so years since Caton's fundamental discovery.

The "Where?" Issue and the Centrencephalic Debate

The first question—regarding the parts of the brain that contributed to consciousness—received partial answers from observations made on human subjects who had localized brain lesions, either from penetrating head wounds during war or from strokes or tumors. But there was a need for care in interpreting these results. For example, a lesion involving one of the occipital lobes could result in blindness for all or part of the visual field on the opposite side, but that did not necessarily mean that it was the occipital lobe that normally brought an image into consciousness: the visual neurons in the occipital lobe could have passed their information on to another, "higher" region of cortex. But there was an even more radical possibility—that the cerebral cortex might not be the site of consciousness at all.

This last idea had been proposed by William Carpenter (1813–1885) at a time when he was Professor of Physiology and Forensic Medicine at University College and Hospital, London, and the author of two textbooks of physiology. Basing his arguments mostly on the anatomy of the brain, Carpenter identified the "sensory ganglia" in the brainstem and thalamus as the sites where conscious sensations were produced (Figure 3.2).[4] The cerebral cortex, although concerned with reasoning and the "will," would perform these last functions unconsciously ("unconscious cerebration"), only bringing them into awareness through actions on the sensory ganglia.

The thalamus was also given a primary role in consciousness by Henry Head (Box 3.1), the pioneering neurologist and general physician at the London Hospital in the capital's impoverished East End. Head had been especially interested in sensation, particularly that of skin and joints, and had been prepared to further his research by having sensory nerves on the back of his own forearm sectioned, so that he and his colleague, William Rivers, could investigate the loss and then the recovery of skin sensation. In his two-volume *Studies in Neurology*, published in 1920, Head wrote: "We believe that the essential organ of the optic thalamus is the centre of consciousness for certain elements of sensation. It responds to all stimuli capable of evoking either pleasure and discomfort, or consciousness of a change of state." Though the cortex was subsidiary, it was nevertheless responsible for supplying precise information about a sensory stimulus to the thalamus. Thus, "the functions of this organ (thalamus) are influenced by the

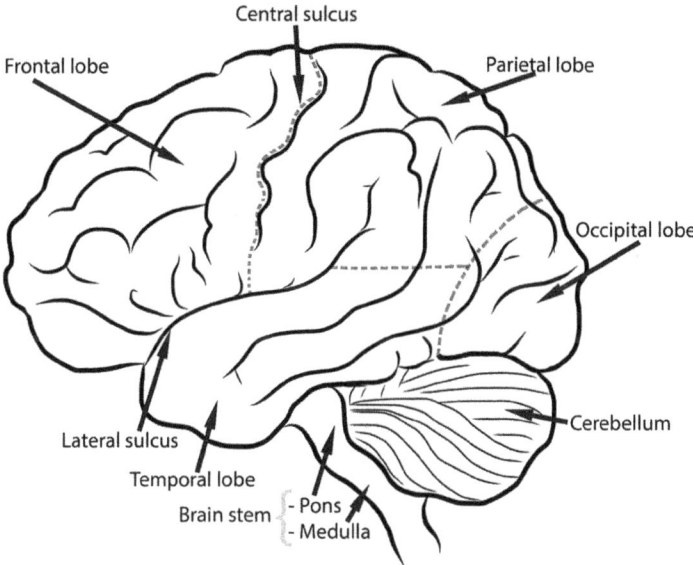

Figure 3.2. View of the outer (lateral) surface of the human left cerebral hemisphere showing the two major clefts (central and lateral sulci), the various lobes, and part of the brain stem (the midbrain is obscured by the temporal lobe). The left thalamus is not seen, being close to the midline above the midbrain, while the precentral gyrus is the cortical fold lying immediately in front of the central sulcus.

Box 3.1 Sir Henry Head (1861–1940)

In an era when British neurology was flourishing under such luminaries as Hughlings Jackson, William Gowers and Gordon Holmes, the most radical thinker within the new specialty was undoubtedly Henry Head. Born into a strongly Quaker family with his father an insurance broker, Henry spent his early years in London. He then boarded at the Charterhouse public school where he became especially interested in biology and physiology, so much so that he travelled to Germany to study under Julius Bernstein at the University of Halle. Returning to Britain, Head entered Trinity College, Cambridge, as a scholar, eventually obtaining first class honors in the Natural Science Tripos. He then went to Europe again, this time spending two years in Prague studying respiration with the noted physiologist Ewald Hering. Back in Britain, Head went from Cambridge to University College Hospital in London, obtaining his medical degree in 1890. Though he would become a superb diagnostician and teacher in general medicine, Head's favorite medical discipline was neurology.

Figure B3.1. Henry Head (left) and William Rivers experimenting in Rivers' room at St John's College, Cambridge. Rivers is testing sensibility on the back of Head's left forearm, in which the cutaneous nerves had been previously cut. With his eyes closed Head is reporting his sensations. (Wellcome Collection)

His first researches in that subject were directed to sensation, beginning with the referral of pain from viscera and continuing with the mapping of human dermatomes by studying the skin lesions of herpes zoster ("shingles"). In 1900 Head became an assistant physician in the London Hospital and his fame as internist, neurologist and teacher continued to grow. The hospital was a busy one, catering to the poor in London's East End and it was there that Head stayed, though he also had a private practice in central London. Busy clinician that he was, Head was also a clinical scientist and deep thinker. Pursuing his early interest in sensation, he chose to have two nerves divided in this left forearm so that he and his Cambridge associate, Dr William Rivers, could examine the return of feeling as the cut nerve fibers regenerated. From this work came the entirely original proposal that skin sensation was mediated by two afferent systems, "protopathic" and "epicritic." The former mediated temperature and pain sensations and was poorly localized, whereas the latter system, evolving subsequently, was concerned with precise touch. Head moved on to investigate pain pathways in the spinal cord and brain. His investigation of patients with thalamic lesions led him to believe that the thalamus was the center for sensory consciousness. Other achievements included an especially thorough study of aphasia and the deduction that the brain created an unconscious model of body posture which he termed the "schema."

Henry Head was not only an outstanding physician and clinical scientist, but a man well versed in the arts. He was a poet, enjoying friendships with Thomas Hardy and Sigfried Sassoon, was authoritative on music and architecture and had numerous other interests. Obliged to retire from medical practice because of parkinsonism, Head maintained a rich intellectual life in the company of friends and his talented literary wife, Ruth.

Source. Jacyna LS. *Medicine & modernism. A biography of Henry Head.* Pittsburgh, PA: University of Pittsburgh Press, 2016.

coincident activity of the cortical centres, and this control is effected by means of paths from the cortex to the thalamus."[5]

Although Carpenter's concept and that of Head may have been well received at the time they appeared, it was the cerebral cortex that was to become tacitly accepted as the site of consciousness. It is uncertain how this happened, though the activities of the neurophysiologists in stimulating and recording from the cortex, the painstaking cytoarchitectural studies of the neuroanatomists, and the continuing description of neurological deficits following cortical lesions, may well have been factors. There was also, lurking in the shadows, the widely held belief that only humans possessed consciousness and that it was the massive development of their cerebral hemispheres that had made this possible. It was against this background that the great centrencephalic controversy arose.

The controversy began when it was the turn of Wilder Penfield (Box 3.2) to deliver the Harvey Lecture to the New York Academy of Medicine in 1936.[6] The annual lectureship was a prestigious one and Penfield a fitting choice as speaker. Then in his middle 40s, Penfield had become one of the world's leading neurosurgeons, as well as the founder of the Montreal Neurological Institute. Much of Penfield's fame had come from his extensive use of electrical stimulation of the exposed human brain; the stimulation was justified in that it enabled

---

Box 3.2 Wilder Penfield (1891–1974)

Penfield spent the first half of his life in the United States, having been born the son of a physician in Spokane, Washington State, and then spending much of his childhood in Wisconsin. An ambitious student, he was to excel academically and in sports after gaining admission to Princeton University. In 1915 he

Figure B3.2. Wilder Penfield. The photograph was taken in 1934, the year that the Montreal Neurological Institute was completed. (Wikimedia Commons)

was awarded a Rhodes Scholarship to study the nervous system under Charles Sherrington at Oxford, but the pressures of World War I prompted him to volunteer as a surgical dresser (assistant) at a military hospital in Paris in 1916. In the following year he came close to death, having been flung high in the air after a German torpedo had blown up the ship in which he was crossing the English Channel. At the end of the war, in 1918, he enrolled at Johns Hopkins Hospital, Baltimore, to finish his medical training, before returning to Oxford to complete his research with Sherrington. Much of this work was neuroanatomical, and was expanded by further studies in Spain with Rio Hortega. Having then taken additional training, this time in neurosurgery with Harvey Cushing, Penfield began his clinical practice in New York. Eager to develop his specialty but encountering obstacles in New York, Penfield accepted an invitation to move to Montreal, where he designed and oversaw the construction of the Montreal Neurological Institute. It was here that Penfield was able to gather a sizeable team, representing many disciplines, to tackle the problems of the nervous system. Much of his own work, performed mostly with Herbert Jasper (Box 3.4), was directed to understanding and treating epilepsy. As part of these studies Penfield and Jasper developed techniques for stimulating the exposed human brain electrically. Not only was Penfield able to create maps of the sensory representation of the body surface and of movements but, by stimulating the temporal lobes, he made important observations on memory. He also proposed a "centrencephalic" hypothesis, in which the higher mental functions were largely executed by the brain stem. Perhaps the most famous neurosurgeon of his time, and certainly the most visible one, Penfield was the recipient of many honors. In his later years his fascination with the neural basis of the mind and his love of medical history prompted him to write several books, in addition to an autobiography.

him to identify function in the different cortical areas, prior to resection of a tumor or an epileptic focus.[7] One of the most examined sites was the precentral gyrus (Figure 3.2), long known to be the primary motor area and able, on stimulation, to produce muscle contractions on the opposite side of the body. What was unexpected, however, was that a patient retained skilled use of the contralateral hand even if Penfield had been obliged to remove the cortical strips immediately behind or in front of the precentral gyrus. To Penfield, this observation suggested that the primary motor area was receiving its instructions not from another region of the cortex but from a deep source—a source where "integration" of neural information was occurring and the will to move might arise. In the 1936 lecture Penfield suggested that "this region lies not in the new brain but in the old—that it lies below the cerebral cortex and above the midbrain." In more detailed lectures, given 15 years later,[8,9] Penfield referred to the region as the "centrencephalic integrating system" and placed it in "certain portions of diencephalon, mid-brain and pons." He also brought forward another argument in favor of consciousness being created subcortically—whereas injury to the brain stem could readily cause loss of consciousness, parts of the cerebral cortex could be removed without any such effect. Moreover, loss of consciousness was a feature of those epileptic seizures thought to arise subcortically.

If 1953 could be considered a high-point for the centrencephalic hypothesis, 1957 was, without doubt, its nadir, for it was in that year that Francis Walshe's devastating rebuttal appeared.[10] Walshe was at that time one of the most senior and distinguished neurologists in Britain's premier neurological institute, the National Hospital for Nervous Diseases in Queen Square, London. Like Penfield, Walshe had studied under Sherrington at Oxford; though the period was a brief one, it was long enough for him to realize that he was unsuited for a career as a neurophysiologist. Instead Walshe had chosen to practice neurology, a specialty in which his penetrating intellect and his literary gifts, allied to a handsome and imposing presence, had assisted his rise to prominence. Some years before his attack on the centrencephalic hypothesis, Walshe had attracted admiration among some of his peers for his skillful demolition of Henry Head's novel concept of "protopathic" and "epicritic" types of cutaneous sensation.[5] In that essay[11] Walshe had not hesitated to employ ridicule, and now, in confronting Penfield, he used it again, summarizing the latter's position in the following way:

> The cerebral cortex, of course, could not be wholly ignored, but it is perhaps not unfair to say that it has been stretched upon the Procrustean bed of a preconceived centrencephalic system, so that we can scarce recognize it.

It is interesting that Walshe spared Head's reputation in his later attack. He may have done so because the older neurologist—by far the more original thinker and a true experimentalist—had died 17 years earlier. Alternatively Walshe may have wished to separate Penfield from intellectual support from another direction. Be as it may, it is not often that a scientist is attacked publicly, and Penfield was both annoyed and dismayed, protesting that he had envisaged the centrencephalic center, while integrating information from various parts of the central nervous system, nevertheless acting in conjunction with the cortex.

However, though it had been badly damaged, the centrencephalic concept did not completely disappear. First, there was the new recognition that dementia could arise in patients whose degenerative changes were largely subcortical.[12] Second, extensive and careful studies of the effects of focal lesions in the rat brain showed that learning in that species depended more on the integrity of the basal ganglia and associated nuclei than on that of the cortex.[13] In addition, there had been greater recognition of the importance of the brain stem following the first published report on the reticular activating system in 1949 (see Chapter 7). But the pendulum of neuroscientific opinion had swung in favor of the cortex being responsible for consciousness, and there it seemed destined to remain.

The Laurentian Symposium, 1953

Meanwhile, in addition to Penfield and Walshe, other neuroscientists and clinicians had been looking into the issue of consciousness and it fell upon Penfield's very able colleague, Herbert Jasper, to bring them together. Jasper chose an inn in the Laurentian mountains of Quebec for the meeting. Since moving to Canada from the United States, Jasper had become an avid skier and the Alpine Inn in St Marguerite was the most prestigious accommodation north of Montreal for winter sports, while providing welcome relief from the heat, humidity and bustle of the city in the summer. Held in August 1953, the symposium was titled "Brain Mechanisms and Consciousness."[14] Jasper restricted the number of participants, knowing that the discussions would be more free and would likely continue over dinner and into the evening (Figure 3.3).

If the group of participants was a small one, it was nevertheless distinguished, its most eminent member being Lord Adrian, the recipient with Sherrington of the Nobel Prize in Medicine or Physiology in 1932. It was Adrian who drew attention to the interaction between spontaneous rhythmic activity in the brain and the responses to different types of stimulus, the relationship between the rhythms and the discharges of individual neurons, and the influence on both of the state of alertness of the human or animal subject.[15] Not surprisingly, given its recent discovery by Giuseppe Moruzzi and Horace Magoun, the reticular activating system was a major topic of discussion at the meeting.

The symposium had been held at a time when the world was still unsettled. True, World War II had ended eight years earlier, but then had come the Korean War, the fighting concluded a bare month before the meeting in the Laurentians. There was a Cold War as well, one which had effectively cut off contact between the Western democracies and the countries of the Communist Bloc. Among its many consequences the Cold War had prevented interactions between the scientists on the two sides and it was those in the communist countries who suffered most. Neuroscientists in the West, having scratched around to build equipment out of World War II surplus, would soon be able to purchase the magnificent amplifiers and oscilloscopes made by Tektronix and Hewlett Packard, in many cases by means of generous support from the United States. It was the start of a glorious era for neuroscience.

Figure 3.3. Participants in the Laurentian Conference on Brain Mechanisms and Consciousness held at the Alpine Inn, Ste. Marguerite, Quebec, in August 1953. Front row, from left: Lashely, Penfield, Adrian, Brazier, Jasper, Bremer, Magoun, Greene. Second row: Gastaut, Rioch, Fessard, Morison, Hess, Olszewski, Grey Walter, Mahut, Jung, Li, Hebb, Kubie. Third row: Ajmone-Marsan, Whitlock, Moruzzi, Nauta, Courtois, Ingvar, Buser. (Reproduced with permission from: Jasper HH. Historical perspective. In: *Consciousness: at the frontiers of neuroscience*, ed HH Jasper et al. Advances in Neurology 77. Philadelphia, PA: Lippincott-Raven, 1998:1–6).

Box 3.3 Sir John Eccles (1903–1997)

In the latter part of a long career Sir John Eccles was widely recognized as the world's foremost central nervous system physiologist. He had been born in Melbourne, Australia, to parents who were both teachers. After completing his schooling Eccles had gone on to study medicine at the University of Melbourne where he had become fascinated by the mind-brain problem. It was a fascination that was to continue throughout his life and that led him, as a Rhodes Scholar, to Britain to study the nervous system under Sherrington at Oxford. It was there, with Sherrington, that he investigated motor units and the excitatory and inhibitory processes underlying spinal reflexes. Having been awarded his D.Phil., Eccles then turned his attention to sympathetic ganglia, becoming a vigorous proponent of electrical transmission at synapses. Returning to Australia in 1937, and in a makeshift neurophysiological laboratory in a Sydney hospital, he attracted Bernard Katz and Stephen Kuffler and worked with them on neuromuscular transmission. It was after World

Figure B3.3. Sir John Eccles, in 1965—two years after his Nobel Prize. (Creative Commons CC BY 4.0)

War II, following moves to the University of Otago (New Zealand) and then to the new Australian National University in Canberra, that Eccles' career blossomed. Employing the new technique of intracellular recording with glass microelectrodes, Eccles was able to reveal the membrane conductance changes responsible for excitation and inhibition at synapses on motoneurons. With a rigorous and comprehensive experimental approach, Eccles went on to investigate other regions of the nervous system (dorsal column nuclei, thalamus, hippocampus and cerebellum), working out the synaptic connectivity in each. During this phase of his career, Eccles attracted neuroscientists from around the world to work with him—Canberra had become the new "Oxford" of neuroscience. His daughter, Rosamund ("Rose"), an excellent neurophysiologist herself, provided valuable assistance during this period. After obligatory retirement from the university, Eccles worked briefly in Chicago and then, more happily and productively, in Buffalo; his last years, during which he continued to write, were spent in Switzerland with his second wife.

The Vatican Symposium, 1964

With neuroscience expanding, the next major symposium on consciousness took place in 1964[16] and had an unusual sponsor—no less a person than Pope Paul VI, the meeting being held within the walls of the Vatican. As with the Laurentian symposium the number of participants was kept small, but those who had been present at the earlier meeting were outnumbered by new faces; all had been selected by the principal organizer, John Eccles (Box 3.3). Eccles, a Nobel prizewinner in the previous year, was a good choice. As a young man intrigued by the mind-brain relationship he had left Melbourne to study with Sherrington in Oxford in the 1920s. After World War II, in the mid-phase of a vigorous scientific career, he had chosen Canberra to create the most productive neurophysiology unit in the world. These qualifications may well have been sufficient for the Vatican but there was also the attraction of having a practicing Catholic as organizer of the meeting.

One of the old guard who came to Rome was Wilder Penfield. Still smarting from the violence of Walshe's unexpected attack on his centrencephalic hypothesis, Penfield had retired from his directorship of the Montreal Neurological Institute in 1960. Though no longer to be seen in the operating theatre, Penfield could be found in his study within the baronial-style building he had lovingly created a quarter-century earlier. Within the high stone walls Penfield had been undisputed master. Under his leadership the Institute had become the foremost clinical neuroscience center in North America, while in Britain its only rival had been London's National Hospital for Nervous Diseases. Now, free of his responsibilities, Penfield had settled down to think and write. Jasper, too, was present at the Vatican symposium. With Penfield no longer the director of the Montreal Neurological Institute, Jasper had decided it was time to move. Busy as ever in the laboratory and fluent in French, Jasper had become the head of the neuroscience research group in the physiology department at the Université de Montréal.

And so the Vatican symposium had begun. Though the Pontiff had asked that no philosophers be invited—the "soul" was to be off-limits—much of the discussion inevitably took a philosophical turn, not least in relation to Lord Adrian's paper. Among the new findings presented were those of Benjamin Libet and Roger Sperry, the former establishing a critical processing time for a conscious perception and the latter describing the unusual behaviors of "split-brain" patients. It was Libet who, in 1983, would publish the results of a yet more important critical timing experiment, one that not only challenged the concept of "free will" but appeared to repudiate it entirely.[17] And in the same year there was another advance for it became possible to visualize a neurotransmitter in the living human brain using positron emission tomography.[18]

Francis Crick (1916–2004) and Functional Brain Imaging

By 1980 some of the most distinguished senior investigators of consciousness were reaching the ends of their careers and Penfield had already died, barely managing to complete his autobiography before succumbing to cancer. But with one major figure gone and others retiring from the study of consciousness, another had appeared—Francis Crick (Figure 3.4). It was a bold move on Crick's part to study consciousness.[19] In doing so, he must have hoped that his objective might yield to the same intelligence, imagination and mental focus that had brought him and James Watson their brilliant success in discovering the double helical structure of DNA a quarter-century earlier.[20] Having gathered younger investigators about him at the Salk Institute, Crick did produce a stream of ideas—and with such a distinguished author, their publication was never a problem. However, it was not difficult, even at the time that they appeared, to be critical of some of the concepts. For instance, how could the prefrontal areas have been the cortical sites enabling the results of visual processing to reach consciousness?[21] As was pointed out at the time, any familiarity with the literature on prefrontal leucotomy should have made such a proposition untenable.[22] Further, neither Crick nor those around him seemed fully aware of the research on consciousness that had already taken place prior to their entry into the field.

In retrospect, there is no doubt that Crick's energy and scientific prominence helped to bring more attention to consciousness as a subject for research. There

Figure 3.4. Francis Crick lecturing. His youthful appearance suggests that the photograph may have been taken not long after his discovery, with Watson, of the double helical structure of DNA. (Wellcome Collection)

was also a new technique available for identifying active regions of the brain, one that was based on the deoxygenation of hemoglobin—functional magnetic resonance imaging (fMRI). It was a seductive technology, not least because it guaranteed a positive result—there would always be one part of the brain with the greatest change. It was hardly surprising, therefore, that the brain became an "open house" for functional imaging, drawing enthusiasts from neuroscience, philosophy and psychology, as well as from the various clinical neurological disciplines. In the great surge of publications that resulted, and continues to this day, there was often a failure to appreciate that the technology was an indirect one for studying the brain's electrical activity. Further, the observed changes were very small, required averaging to be detected and were often not reproducible. Even worse, it later transpired that the underlying mathematical algorithms for the computer analysis were flawed in many of the studies.[23] Nevertheless, with a burgeoning interest in the brain, and with fMRI a ready if somewhat suspect tool, it was unsurprising that the rate of publication on various aspects of consciousness accelerated.

Montreal Symposium, 1997

There were more meetings too. Of the latter, the most significant may have been one in Montreal in 1997,[24] a meeting that honored the 90th birthday of one of the great pioneers of neurophysiology—Herbert Jasper (Box 3.4), the same person

Box 3.4 Herbert H. Jasper (1906–1999)

The small man with the trim moustache and spectacles might have been mistaken for an office worker. Yet, among the thousands of scientists who have studied the brain, it is doubtful if any—with the possible exception of Edgar Adrian—could have matched the breadth of experience of Herbert Jasper. Born in Oregon, USA, the son of a minister, Jasper's higher education began in that state's Williamette University, where he studied philosophy and, at the same time, developed an interest in neuroscience and psychiatry. Later studies at the University of Iowa gave him his first experience of working directly with nerve and muscle and this led him to France and enrollment as a doctoral student at

Figure B3.4. Ninety-year-old Herbert Jasper (left) talking to David Hubel at the time of the 1997 symposium on consciousness in Montreal. Fifty years earlier they had been medical students together. (Photo by author)

the Sorbonne; while in Paris he worked on crustacean neuromuscular systems in the department of Louis Lapique. Then followed an appointment at Brown University in Providence, Rhode Island, and the start of investigations into the EEG, following key publications by Hans Berger in Germany and by Adrian and Matthews in the UK. With a newly built machine, and with increasing experience, Jasper was able to convince Penfield that the EEG could be used to localize epileptic foci in the brain. Impressed by the results, and by Jasper's subsequent willingness to drive to Montreal to make intraoperative recordings, Penfield invited Jasper to join him at the Montreal Neurological Institute. With new laboratories created for him, Jasper was able to broaden his activities, not only developing EEG but attempting correlations between EEG abnormalities and subsequent pathological findings. In the middle of all this activity Jasper enrolled in the McGill MD program, often depending on David Hubel, then a young student, to provide him with lecture notes.

It was in the post-war years that Jasper's career reached its height, with numerous young and established neuroscientists coming to work in Montreal with him. Among his many achievements was the identification, with Allan Elliott, of GABA as the major inhibitory transmitter in the mammalian brain. On the clinical front he had assisted Penfield in his brain mapping observations, and had developed the 10-20 electrode system as the standard arrangement for human EEG recordings. The use of monopolar needle electrodes for intramuscular EMG studies was another innovation. Jasper's creativity, especially in the field of neurotransmitters, continued following his move to the University of Montreal. Despite early parkinsonism, Jasper's mind remained active into his nineties, as he proved by delivering two excellent lectures at the 1997 Montreal meeting in his honor. Among his many honors was the Albert Einstein World Science Award (1995).

who had organized the Laurentian symposium on consciousness 44 years earlier. Not only did Jasper give a historical introduction—who better to have done it?—but he described his work on the thalamic recruiting system, his recordings from the human thalamus during stereotaxic neurosurgery, and the EEG as consciousness was lost during epileptic seizures.[25] At the same meeting the youthful philosopher David Chalmers drew attention to the difficulty of solving the "hard" problem; for example, even if all the neural activity underlying sensory processing were known, it would still not explain how this resulted in a conscious perception.[26] What was it that translated nerve impulses into "blueness" or "redness," the sound of a C major chord, or the scent of a rose? Also present at the meeting was David Hubel, first known to Jasper as a medical student but long since the foremost investigator of the visual system and a Nobel Laureate.

Leaving aside the tantalizing "hard" problem, most of the pieces essential for an understanding of the neural mechanisms of consciousness were now in place.

Most, but not all. There was one more key observation to be reported, and this was the experimental confirmation by Rodrigo Quiroga and his colleagues in 2005 that there were such entities as "grandmother cells" in the human cortex.[27] While there still remained uncertainty over the full functions of such brain structures as the basal ganglia and cerebellum, enough was now known for a plausible account of the neural underpinnings of consciousness. And it was one of those accounts that came with a surprise, for it placed consciousness not in the cortex, where it was generally held to arise, but in the brain stem.

Back to the Centrencephalon!

Antonio Damasio (1944; Figure 3.5) had been born in Lisbon, the capital of Portugal, and had studied medicine there before moving to Boston for doctoral research on

Figure 3.5. Antonio Damasio autographing while at a conference in São Paulo, 2013. (author: Fronteiras do Pensamento; Wikimedia Commons CC-SA 2.0)

aphasias with Norman Geschwind. Next had come an appointment as associate professor of neurology at the University of Iowa and it was while in Iowa that the first of several notable books on consciousness appeared, one that took up the scientific implications of the remarkable story of the brain-injured railroad engineer, Phineas Gage.[28] However, it was Damasio's interest in the nervous systems of different animal species and in the place of feeling and emotion in cognition that led him to suggest, in a later book, that a basic consciousness had evolved in the brain stem, specifically in the dorsal region of the midbrain and upper pons (Figure 3.2).[29] It was a bold speculation but it was one that had a history, for had not Penfield identified the brain stem as the seat of consciousness fifty years earlier—and been punished for it? And a full century before Penfield, there had been William Carpenter who, in his popular textbook of physiology, had reached a similar conclusion.

The wheel had come full circle.

With the historical outline having been given and some of the major figures identified, the next step in this journey is to enquire who, other than we human beings, might also be conscious.

Notes and References

1. Cited by Blakemore in his 2005 Harveian Oration: Blakemore C. In celebration of cerebration. *Clinical Medicine.* 2005; 5: 589–613.
2. Willis T. *Cerebri Anatome: cui Accessit Nervorum Descriptio et Usus.* London, England: J. Flesher, 1664. Cited by Blakemore C., 2005.
3. Caton R. Electrical currents in the brain. *British Medical Journal.* 1875; 2: 278.
4. Carpenter WB. *Principles of human physiology.* 4th ed. London, England: Churchill, 1853 Cited by: Walshe FMR, 1957, noted below.
5. Head H. *Studies in neurology.* Vol. 1–2. London, England: Henry Frowde, 1920.
6. Penfield W. The cerebral cortex in man. I. The cerebral cortex and consciousness. *Archives of Neurology and Psychiatry.* 1938; 40: 417–442.
7. Penfield was not the first to stimulate the exposed human cortex electrically during life; Roberts Bartholow, an American surgeon, had, in 1874, reported doing this. See: Zago S, Ferrucci R, Fregni F, Priori A. Bartholow, Sciamanna, Alberti: Pioneers in the electrical stimulation of the exposed human cerebral cortex. *Neuroscientist.* 2008; 14: 521–528.
8. Penfield W. Mechanisms of voluntary movement. *Brain.* 1954; 77: 1–17.
9. Penfield W. Studies of the cerebral cortex of man: a review and an interpretation. In: Delafresnaye JF, ed. *Brain mechanisms and consciousness.* Springfield, IL: Charles C. Thomas, 1954:284–304. .
10. Walshe FMR. The brain-stem conceived as the "highest level" of function in the nervous system; with particular reference to the "automatic apparatus" of Carpenter (1950) and to the "centrencephalic integrating system" of Penfield. *Brain.* 1957; 80: 510–539.

11. Walshe FMR. The anatomy and physiology of cutaneous sensibility: a critical review. *Brain.* 1942; 65: 48–112. This article, and the background that led to it, is discussed in a fascinating historical account by Alistair Compston, a former editor of *Brain*. (Compston A. From the archives. *Brain.* 2011; 134: 920–923). Head's concepts subsequently found validation as part of the background to Melzack and Wall's "gate" hypothesis of pain (Melzack R, Wall PD. Pain mechanisms: a new theory. *Science.* 1965; 150(3699): 971–979).
12. Brown RG, Marsden CD. Subcortical dementia: the neuropsychological evidence. *Neuroscience.* 1988; 25: 363–387.
13. Thompson R, Crinella FM, Yu J. *Brain mechanisms in problem solving and intelligence: a lesion survey in the rat brain.* New York, NY: Plenum Press, 1990.
14. Delafresnaye JF, ed. *Brain mechanisms and consciousness.* Springfield, IL: Charles C. Thomas, 1954.
15. Adrian ED. The physiological basis of perception. In: Delafresnaye JF, ed. *Brain mechanisms and consciousness.* Springfield, IL: Charles C. Thomas, 1954:237–243.
16. Eccles JC, ed. *Brain and conscious experience.* New York, NY: Springer-Verlag, 1966.
17. Libet B, Wright EW Jr, Feinstein B, Pearl DK. Time of conscious intention to act in relation to onset of cerebral activities (readiness potential): the unconscious initiation of a freely voluntary act. *Brain.* 1983; 102: 623–642.
18. Garnett ES, Firnau G, Nahmias C. Dopamine visualized in the basal ganglia of living Man. *Nature.* 1983; 305: 137–138.
19. Crick FHC. Thinking about the brain. *Scientific American.* 1979; 241: 219–232.
20. Watson JD, Crick FHC. A structure for desoxyribose nucleic acid. *Nature.* 1953; 171: 737–738.
21. Crick FHC, Koch C. Are we aware of neural activity in primary visual cortex? *Nature.* 1995; 375: 121–124.
22. McComas AJ. Discussion following presentation: Koch C. The neuroanatomy of visual consciousness. In Jasper HH, Descarries L, Castellucci VF, Rossignol S, eds. *Consciousness: at the frontiers of neuroscience.* Advances in Neurology Vol. 77. Philadelphia, PA: Lippincott-Raven, 1998:229–243.
23. Eklund A, Nichols TE, Knutsson H. Cluster failure: why fMRI inferences for spatial extent have inflated false-positive results. *Proceedings of the National Academy of Sciences of the United States of America.* 2016; 113 (28): 7900–7905.
24. Jasper HH, Descarries L, Castellucci VF, Rossignol S, eds. *Consciousness: at the frontiers of neuroscience.* Advances in Neurology Vol. 77. Philadelphia, PA: Lippincott-Raven, 1998.
25. Jasper HH. Sensory information and conscious experience. *Advances in Neurology.* 1998; 77: 33–48.
26. Chalmers DJ. The problems of consciousness. *Advances in Neurology.* 1998; 77: 7–16.
27. Quiroga RQ, Reddy L, Kreiman G, Koch C, Fried I. Invariant visual representation by single neurons in the human brain. *Nature.* 2005; 435: 1102–1107.
28. Damasio A. *Descartes' error: emotion, reason and the human brain.* New York, NY: Putnam, 1994.
29. Damasio A. *Self comes to mind.* New York, NY: Random House, 2012.

4
Who Else Is Conscious?

Historical Attitudes

> And God said, Let us make man in our image, after our likeness: and let them have dominion over the fish of the sea, and over the fowl of the air, and over the cattle, and over all the earth, and over every creeping thing that creepeth upon the earth.
>
> Genesis 1:26

And, indeed, in the millennia that followed the biblical exhortation, Western societies proceeded to exercise their God-given dominion. Not only were animals hunted or farmed to provide sources of food and clothing—an arguably legitimate relationship between human and nonhuman species—but animals found themselves employed as beasts of burden; as haulers of carts, carriages, and ploughs; and even as vehicles of war. While some were favored by adoption as pets, more exotic species found themselves objects of curiosity in zoos and aquaria, and, more recently, others provided material for study by physiologists and pharmaceutical companies.[1]

Putting aside the Bible and other religious writings, for those who reflected upon the ethics of human–animal relationships, there was a comforting thought—that only humans were conscious and had minds, and this by virtue of their large brains and by the development of the cerebral cortex in particular. Even now, despite advances in neuroscience and in the study of animal behavior (cognitive ethology), this is probably the view that most of the population, an educated population at that, would subscribe to. In the remainder of this chapter, it will be argued to the contrary, that not only is there a compelling case for consciousness in animals but that the latter provides valuable clues for an understanding of how consciousness evolved in the animal kingdom.

Inevitably, one begins with Charles Darwin.

Darwin and *The Descent of Man*

In 1871 Charles Darwin (Box 4.1) published one of his great works, *The Descent of Man*.[2] The first major book, *The Voyage of the Beagle*,[3] had appeared in 1839 and *The Origin of Species*[4] had followed 20 years later. If Darwin's greatest intellectual concern throughout his life was the origin and nature of "man," it was a

---

Box 4.1 Charles Darwin (1809–1892)

One of the most important scientists who ever lived, Charles Robert Darwin was born in Shrewsbury, United Kingdom, the son of a doctor and the grandson of Erasmus Darwin, the noted doctor and philosopher. Although

Figure B4.1. Charles Darwin, age 40. As a young man, prior to *Beagle*, Darwin had enjoyed collecting insects and shooting. (Lithograph by T.H. Maguire, 1849; Wellcome Collection)

Charles had entertained medicine for a career and had enrolled at the University of Edinburgh medical school, he soon became disenchanted with his studies and far more interested in natural history. His fascination with the latter deepened after giving up medicine for an arts degree at the University of Cambridge; during this time he came under the influence of John Henslow, professor of botany, and it was Henslow who suggested Darwin serve as a gentleman-naturalist in a forthcoming naval expedition. The purpose of the expedition was to chart the South American coastline; the ship involved, the *HMS Beagle*, was to be commanded by Captain FitzRoy. During the voyage Darwin studied marine invertebrates but made his most important observations on land, where he ventured far on horseback in the company of a single assistant. In addition to describing and collecting local fauna, he discovered new types of fossil; as an amateur geologist, he recognized raised beaches as irrefutable evidence that the seafloor had been lifted in the distant past. While visiting the Galapagos Islands Darwin discovered a dozen or so new species of finch, as well as various types of tortoise, and noted that the individual species were peculiar to particular islands. These observations ultimately led him to conclude that one species could evolve into another, having developed variations in structure that would prove advantageous. This view and its corollary, that existing life-forms were descended from more lowly creatures, were set out with exhaustive evidence in *The Origin of Species* many years later. The importance of Darwin's discovery (simultaneously with that of Alfred Wallace) was recognized immediately; that it was not universally accepted was due to it running counter to religious belief in a Creator.

Darwin, having retired to his home in Kent, continued to collect and observe and to write books and papers. His public appearances became fewer, however, in part because he was plagued with chronic ill-health. Regarded in his own lifetime as a scientific giant, Darwin was buried in Westminster Abbey.

concern that he had been reluctant to emphasize in *The Origin of Species*—and for good reason, given the hostility toward such an idea not only from the very powerful Church of England in his own country but from many well-educated but nevertheless skeptical members of the public. With the passage of years, however, Darwin's views had come to be supported by other notable scientist-philosophers, the most prominent of whom had been Thomas Henry Huxley (Box 2.2)—it was Huxley who first reviewed *The Origin of Species*, having done so for Britain's leading newspaper, *The Times*. Yet, despite the growing approval and support of many contemporaries, Darwin was not satisfied. Working in the quiet of Down House, his home in Kent, United Kingdom, he had kept notes on the behavior of different animal species and of primitive human societies and had maintained an extensive correspondence on these matters with authorities in Britain and Europe. This huge body of information, the thoughts and musings of many years, the observations that he and others had made, and, above all, his conviction that humans had evolved from more lowly species—all this needed to be set out in print. Moreover, Darwin was concerned about the further evolution of humankind, a topic that followed naturally from his observations on sexual selection in animals. Marriage, he thought, should not be entertained by the poor, for fear of condemning future generations to similar lives of poverty. And so, with this injunction given in its final pages, *The Descent of Man* appeared in 1871. Just as *The Origin of Species* had been 12 years earlier, it was an instant success.

It was the third chapter in *The Descent of Man* that proved the most relevant to the issue of animal consciousness. In it, Darwin pointed out that

> lower animals, like man, manifestly feel pleasure and pain, happiness and misery. Happiness is never better exhibited than by young animals, such as puppies, kittens, lambs etc., when playing together.

Darwin went on to describe other human-like traits in animals, such as courage, timidity, and ill-temper, giving examples of each. At the end of the fourth chapter, the following statement appeared as a summation of the properties of animal "minds":

> Nevertheless the difference in mind between man and the higher animals, great as it is, certainly is one of degree and not of kind. We have seen that the senses and intuitions, the various emotions and faculties, such as love, memory, attention, curiosity, imitation, reason etc., of which man boasts, may be found in an incipient, or even in a well-developed condition, in the lower animals.

Given his preoccupation with evolutionary forces—in plants as well as in animals—it was perhaps not surprising that Darwin should have wondered at what point consciousness had appeared in the animal lineage.[5] Though it was a question that would vex him throughout the latter part of his life, it was not for want of observations, many of which he jotted down in a notebook. As an example he argued that even a humble earthworm must possess some kind of mind, since he had noticed that its normal sudden withdrawal on exposure to light would not occur if the worm was engaged in some task.

Darwin, who carried on an extensive correspondence and read widely, was far from alone in his quest for the origins of consciousness. Indeed one of his contemporaries, W. K. Clifford, made two interesting suggestions, one of them for a fundamental "mind-stuff" and the other for cognition arising after the evolution of sensory experience ("sentience"). Thus:

> Mind-stuff is the reality which we perceive as matter. A moving molecule of inorganic matter does not possess mind or consciousness, but it possesses a small piece of mind-stuff. When molecules are so combined together to form the film on the underside of a jelly dish, the elements of mind-stuff which go along with them are so combined as to form the faint beginnings of Sentience. When the molecules are so combined as to form the brain and nervous system of a vertebrate, the corresponding elements of mind-stuff are so combined as to form some kind of consciousness. [6]

This was a remarkable proposition for the time, and one that clearly anticipated the thinking of contemporary philosophers.[7]

T. H. Huxley on Consciousness in Animals

Three years after the appearance of *The Descent of Man* it was the turn of Darwin's friend and champion, Thomas Huxley, to express an opinion on animal consciousness, and this he did in an invited lecture delivered in Belfast. Titled *On the Hypothesis that Animals are Automata, and its History*,[8] Huxley drew on the reasoning of Descartes and, as a gifted physiologist himself, was able to make use of observations on behavior in frogs and humans with damage to their respective nervous systems. In his carefully argued discourse, Huxley concluded that animals ("brutes"), though acting largely on instinct, were nevertheless conscious:

> We know, further, that the lower animals possess, though less developed, that part of the brain which we have every reason to believe to be the organ of consciousness in man; and as, in other cases, function and organ are proportional, so we have a right to conclude it is with the brain. And the brutes, though they may not possess our intensity of consciousness, and though, from the absence of language, they can have no trains of thoughts, but only trains of feelings, yet have a consciousness which, more or less distinctly, foreshadows our own.

While attributing consciousness to animals, Huxley was surely mistaken in his remarks about "absence of language" and, by implication, "no trains of thoughts." Animals, including insects and birds, certainly have language, both vocal and signed, and the challenge for human observers is to identify and interpret the nuances in each species. Further, it is likely that, using their language, animals have trains of thought—as evidenced by their ability to solve new problems. Interestingly, human subjects born deaf but subsequently familiar with American Sign Language experience streams of consciousness with a succession of internally visualized signs (as opposed to having an inner "auditory" voice).

Problem-Solving in Animals

To both Darwin and Huxley it was apparent that animal behaviors indicative of emotion (happiness, fear, rage, jealousy) could only be experienced by a conscious brain. Unlike Huxley, however, Darwin went further by recognizing other animal activities that seemed to have a cognitive element—social interactions, attention, memory, imagination, and reason. Indeed, it is in the solution of novel problems that the most convincing arguments for animal consciousness are to be found, and the person who was the greatest advocate of this view was Donald Griffin (Box 4.2).

Box 4.2 Donald Griffin (1915–2003)

Donald Griffin was a neuroscientist, naturalist, and zoologist whose reputation deserves to increase as the presence of consciousness in animals comes to be more widely accepted. Griffin was born in Southampton, New York, where his father was self-employed as a historian and novelist; in his boyhood Griffin was also influenced by a maternal uncle who was a professor of biology at Harvard. At a young age Griffin was already trapping and skinning small mammals and, while in his teens, began banding bats. After entering Harvard as an undergraduate in 1934, Griffin's continuing interest in bats led to him showing, for the first time, that these creatures emitted ultrasonic cries while in flight. Not long after, Griffin was able to demonstrate, with fellow graduate student Robert Galambos, that bats used the reflections of these high-frequency utterances for their navigation—the first description of echolocation and an early account of a highly unusual sensory adaptation in an animal. Griffin's other major interest was in bird navigation, and this formed the subject of his graduate thesis; together with bat echolocation, it continued to occupy him for the remainder of his life. Among the many new methods of study that he devised, Griffin took up aviation so that he could follow the flights of individual birds for hours.

After occupying a faculty position at Cornell University for some years, Griffin was invited back to Harvard as head of zoology in 1953. Some 12 years later, however, he accepted another invitation, this time to organize and direct Rockefeller University's new Institute for Research in Animal Behavior, and he remained in this position until 1986.

Almost to the day that he died, in his 88th year, Donald Griffin continued to experiment and observe, while remaining a major champion for animal consciousness.

Source. Griffin DR. In: *The history of neuroscience in autobiography*, ed LR Squire, vol 2. San Diego, CA: Academic Press, 2014: 67–92.

An indifferent student, Griffin had nevertheless become intensely interested in biology and, after taking part in bird navigation experiments as an amateur "bander," had become fascinated by bats. While still in his 20s he co-authored a paper on that species that gave the first description of bat navigation by means of echo-location.[9] Despite his success as an experimental biologist, and his stature as head of the Zoology Department at Harvard University, Griffin would discover that his first major publications on animal intelligence—*The Question of Animal Awareness* (1976)[10] and *Animal Thinking* (1985)[11]—would be poorly received by the majority of his scientific peers, perhaps in part because of the lack of attention given to animal consciousness following Darwin's death and that of his younger colleague, George Romanes (1848–1894).

Undeterred, Griffin continued to act as a lonely advocate for the existence of consciousness in animals, *Animal Minds* appearing in 1992.[12] There was, however, one encouraging development during this period, this being the award of the Nobel Prize in Physiology or Medicine in 1973. Unusually, the Nobel Committee had looked beyond the conventional boundaries of medicine and physiology and had decided to honor three scientists who had devoted their lives to the study of animal behavior. The surprised recipients of the Prize were Konrad Lorenz, Nikolaas Tinbergen, and Karl von Frisch. Among their many achievements, Lorenz had attracted attention for his description of "imprinting" in newly born animals, while Tinbergen had become renowned for his convincing experimental designs; however, it was von Frisch's discovery, that a honey bee performed a precise dance to inform other members of the hive of the whereabouts of a source of nectar, that had the greatest impact on the imagination of the public. Darwin would surely have nodded approvingly when, in their Nobel acceptance speeches, each recipient sought to justify his award by pointing out that much might be learned about abnormal human conduct through the study of animal behavior.

But back to Griffin—what were some of the behaviors that had caused him to postulate that animals could *think*? And what other examples might have been added to Griffin's list? The following is a somewhat arbitrary collection, the source material coming from Griffin and other authors, who have been cited wherever possible. Inevitably, many of the accounts are anecdotal and therefore have no hard evidence to support them. On the other hand, the similarities of some of the accounts add credibility, and there can be no gainsaying the authenticity of behaviors captured by camera and displayed on the Internet.[13]

Parrots

The great advantage of using parrots to study behavior is that these remarkable creatures are unique among animals in being able to utter human words and sentences and thus, at least in theory, inform the observer directly of their thoughts and observations. One of the most studied parrots has been an African Grey called "Alex," the property of Irene Pepperberg. Alex was able to identify colors and small numbers and could name up to 40 different objects in English, some of which he would request as playthings. If given the wrong object, Alex would utter "No!" in protest.[14] Alex's last words to his owner, uttered prior to the parrot's death in 2007, were "You be good. See you tomorrow. I love you."

In *The Descent of Man* Charles Darwin cites another example of prowess with words by a parrot:

> Admiral Sir B.J. Sullivan, whom I know to be a careful observer, assures me that an African parrot, long kept in his father's house, invariably called certain persons of the household, as well as visitors, by their names. He said "good morning" to every one at breakfast, and "good night" to each as they left the room at night, and never reversed these salutations. To Sir B.J. Sullivan's father, he used to add to the "good morning" a short sentence, which was never once repeated after his father's death. He scolded violently a strange dog which came into the room through the open window, and he scolded another parrot (saying "you naughty polly") which had got out of its cage and was eating apples on the kitchen table.

Among the recent offerings on the Internet is Petra, an African Grey that has been caught conversing with Amazon's home assistant, Alexa, and giving the latter instructions for shopping and for dimming the lights in the home. Equally impressive, though for an entirely different type of behavior, is a cockatoo, Snowball, that dances—very well—to music by the Backstreet Boys!

Crows

Crows and other members of their family (ravens, rooks, jackdaws, and jays) have long had a reputation for intelligence. For several months at least, these birds remain capable of recognizing a person who may have treated them badly. When presented with a worm wriggling just out of reach in a glass jar half-filled with water, a crow has been observed, just as in the well-known fable by Aesop, to raise the water level by picking up stones with its beak and dropping them in the jar until the worm could be seized (Figure 4.1). Another crow was seen to free a stick by pulling on a string, then using that stick to obtain a longer one, and finally using the longer stick to retrieve a food item from inside a cage. In each of these examples, the crow did not act immediately but only after a considerable pause—as if the bird had been working out the solution to the problem in its head. As a final example of crow intelligence, birds in Japan have been observed to obtain food from nuts and shells by dropping the objects on to the road and then reclaiming the contents after the shells had been run over and broken by vehicles. What is even more remarkable, however, is that the Japanese crows have learned to restrict this activity to pedestrian crossings, waiting patiently with humans for the traffic to stop before venturing forth.

Figure 4.1.

Dogs and Cats

If parrots and crows are the cognitive champions of the avian world, their canine counterparts are border collies and poodles. Collies, of course, are bred for sheep-herding, a skill dependent on their natural intelligence and their ability to interpret commands whistled by the shepherd. One border collie, Chaser, was shown two new stuffed toys, balls, or other objects a day by its owner, retired psychology professor John Pilley. At the end of 3 years Chaser was able to identify over a thousand such objects. And, indeed, anyone who has had a dog or cat as a pet can testify to the ability of their charge not only to obey a range of commands and to quickly learn tricks for a reward but, more impressively, to interpret the behaviors of their masters. The putting on of a coat or the picking up of a leash, for example, become clear signals that one is about to be taken for a walk, while the opening of a refrigerator heralds the next feed. On the other hand, bringing a stick to an owner, with a barked command that it should be thrown for retrieval, is an obvious example of human training by a dog!

Octopus

The ability of an octopus[15] to somehow squeeze, contort, and elongate its body in order to escape through a small hole is a source of wonder to a human observer. Even more astonishing was the mystery of the disappearing lump-fish in a marine laboratory. Eventually the staff discovered that an octopus was responsible—that, after the staff had left, it was able to climb out of its tank into an adjacent one, seize and devour a lump-fish, and return to its own tank.[16]

Another example of octopus intelligence is that given in an interview with Jerome Lettvin. Referring to an experience witnessed with his son Jonathan in the Zoological Station in Naples, Italy, he recounted:

> I teased JD—he was a big octopus, he had a 5 foot spread of arms—I teased him by holding a fish down for him to grab, then pulling it back. JD would start going black-white-black-white indicating a high degree of irritation. Then JD decided to play a joke on me. Let me tell you, anybody who says octopuses don't have a sense of humor . . . forget it; they're good!
>
> The next morning I walk into the lab and JD is up on the edge of his 20 foot tank. I walk in and smack! . . . right in my face . . . he let out a huge squirt of water! He had been waiting for me to come in. Jonathan is in stitches, and then Jonathan says, "Take a look!" and I turn around and there are splotches of water all around the region of where my head would be when I come in. The octopus had planned his revenge and he had been practicing ahead of time. That sounds like a ridiculous story, eh? Well, it isn't.[17]

Nonhuman Primates

Charles Sherrington, during his time as professor of Physiology at the University of Liverpool in the United Kingdom, once decided on a whim to peer back into the animal room that he had just left. Applying his eye to the keyhole of the door, he was startled to see the eye of a gorilla looking back at him.[18]

Charles Darwin, in *The Descent of Man*, tells the story of a baboon revenging itself on a human tormentor:

> At the Cape of Good Hope an officer had often plagued a certain baboon, and the animal, seeing him approaching one Sunday for parade, poured water into a hole and hastily made some thick mud, which he skilfully dashed over the officer as he passed by, to the amusement of many bystanders. For long afterwards the baboon rejoiced and triumphed whenever he saw his victim.

In an attempt to communicate with a chimpanzee, Allen and Beatrice Gardner (University of Nevada) employed American Sign Language; their rationale was that chimpanzees, like other nonhuman primates, had restricted ability with their vocal apparatus and that it was therefore illogical to use spoken words.[19] After 4 years of training beginning in infancy, their chimpanzee, Washoe, had mastered more than 130 signs and could, for example, use these correctly to identify pictures of objects. On one occasion, having been presented for the first time with a picture of a swan, Washoe signed "water bird." Following the Gardners, other investigators have used various modifications of the signing protocol and have also obtained persuasive examples of communication with chimpanzees and apes.

Perhaps the most astonishing example of intelligent behavior in a chimpanzee is that displayed by a Japanese animal named Ayumu at Kyoto University. Given free access to a computer from an early age, Ayumu proved capable of memorizing the positions of the digits 1 to 9 that were momentarily flashed on to a touchscreen and of then rapidly touching them in the correct numerical order; no human has been able to match Ayumu's skill.[20,21]

Before leaving the subject of the abilities of these primates, it should be added that they—like birds—are capable of finding or making elementary tools from twigs or plant stems in the search for food.

Dolphins

Both dolphins and whales have large brains with convoluted surfaces. While it is difficult to test the intelligence of whales, it is known that they normally "sing" to each other. The mental skills of dolphins are better known and have been frequently displayed for the amusement of the public in aquaria of the "Sea World" type. The dolphins can be taught, on command, to play with balls, paddle backwards while remaining upright, leap through hoops above the water, and so on. So quick are the dolphins to master new "tricks" that a major problem confronting the trainers is to find novel ways of overcoming boredom in their animals. In the wild, dolphins have developed successful strategies for avoiding entrapment in nets intended for tuna and also for hunting fish as a team.

Further credence to the intelligence of dolphins has come from analysis of their vocal sound patterns. Thus researchers at the Karadag Nature Preserve in Russia observed that each of a pair of dolphins appeared to communicate with the other in "sentences;" each sentence was formed by a sequence of critically timed clicks, the latter varying in their sound components. When one dolphin spoke in this way, the other remained silent and appeared to be listening, only replying when the first one had finished.[22] Researchers in Florida have documented the ability of one dolphin to instruct another, through speech, in the opening of a cylinder containing a fish.

While the dolphin may provide the most sophisticated example, speech and other methods of communication between members of the same species are, of

Figure 4.2. Assassin bug with distant termite mound.

course, prevalent in the animal kingdom and, as such, provide further evidence strongly suggestive of cognition.

But What About . . . Insects?

The very effective waggle-dance employed by honeybees to indicate a source of nectar has already been noted. While this behavior is evidently genetically programmed, it is not stereotyped, for the nectar may be near or far, and at any angle to the sun's rays, and this information must be conveyed correctly by the bee on its return to the nest or hive.

An equally remarkable example of complex behavior in an insect is that of a species of assassin bug in tropical rain forests (Figure 4.2).[23] This bug feeds on termites but, in order to do so, must undertake precautions not to become prey itself. The bug therefore coats itself with pieces of termite nest so that it exudes an odor familiar to any investigating soldier termites. The assassin bug then reaches into the nest and seizes a worker termite, killing the latter and sucking out the internal organs. The bug then uses the dead termite as bait, pushing the exoskeleton into the opening of the nest, where it is grasped by a worker. Next, the dead termite and the live one attached to it are withdrawn and the live one eaten by the assassin bug. The fresh exoskeleton is then used as bait for another "fishing" expedition in the nest, and the whole process can be repeated many times.

Does the complexity of this behavior, and that of the honeybees, with its adaptability to the exigent circumstances, denote consciousness on the part of the insect? It is, of course, impossible to answer this question and will inevitably remain so. However, for an unprejudiced extraterrestrial observer, one making comparisons of behaviors in the various life-forms on Earth, including humans, the answer would probably be "yes." And, as an aside, this line of reasoning, embodied in the well-known Turing test, is that used to judge the presence or absence of consciousness in computer-driven machines—as discussed in Chapter 14.

Related to the previous question is the nature of the mental experience in those life-forms that, for the reasons given earlier, almost certainly *do* have consciousness. The issue was brought to the fore by the philosopher Thomas Nagel in an article titled "What Is It Like To Be a Bat?"[24] However, though the bat, because of its poor vision and reliance on echolocation, may be a difficult instance, other species offer a glimmer of hope. In the case of chimpanzees and other nonhuman primates, it is likely that their perceptual consciousness is very similar to our own, given their discriminative abilities and the similar physiological systems for vision, hearing, and touch. In contrast, the "inner voice" that is so well developed in humans, as a consequence of the extensive and varied use of language in speech and writing, would be a smaller part of a chimpanzee's consciousness. Though it is a much bigger jump, we can have an inkling of the nature of a rat's consciousness too, this time from the responsiveness of its hippocampal neurons, as observed in microelectrode recordings. Such studies indicate that, in addition to odors, some of the most important features of a rat's life are the places it encounters. But as we move from mammals into other branches of the animal kingdom, so the guesses about the nature of mental experience inevitably become wilder. Perhaps, following Nagel, we should not even make the attempt.[25]

Availability of Neurons for Consciousness

In Chapters 9 and 12 evidence is brought forward to suggest that the recognition of a person or object depends on the activation of particular "gnostic units" ("grandmother cells") and suggestions are made for the numbers of neurons comprising such units. However, the number of conscious experiences possible in a member of any given species will depend not only on the number of neurons involved in a single experience (gnostic unit) but on the magnitude of the neuron population available. A comparison of brain sizes and shapes in different animal species (Figure 4.3) indicates that the range of neuron populations is likely to be large, and this is borne out in Table 4.1, which gives the total numbers of neurons in the nervous systems in a variety of species. The table contains a number of

WHO ELSE IS CONSCIOUS? 45

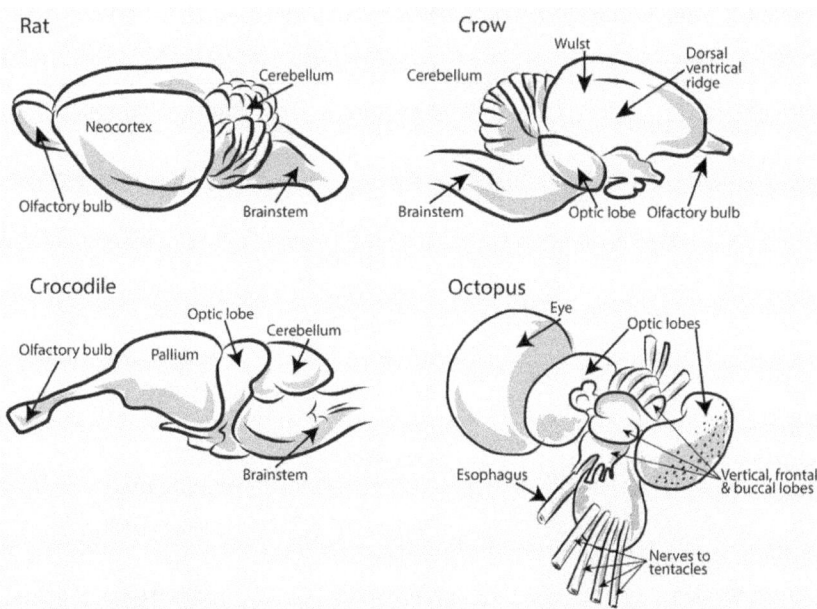

Figure 4.3. Varied gross structure of four animal brains (dissimilar scales). In the crow brain, the "wulst" is a raised area largely devoted to vision. Only the rat brain has a neocortex with cell columns. The octopus has a nerve to each of its eight tentacles (four nerves not shown).

surprises. For example, an insect as small as an ant, or the even smaller fruit fly, has as many as a quarter million neurons. Moving up, a raven, by virtue of the high packing density of neurons in its brain, contains 3 times as many neurons as a larger creature, a cat. At the high end of values, an African elephant has 3 times as many neurons as a human, though only half as many in its cerebral cortex. However, of those species examined, it is not humans that have the greatest number of cortical neurons but dolphins—and by a very considerable margin.

As important as are the numbers of neurons for considering the possibility of consciousness in animals, there is another body of information that has accumulated through decades of neuroanatomical and neurophysiological studies. These investigations have repeatedly shown that key neural circuits—for special senses, movement, and emotions—are present not only in humans but also in the other mammals examined, though sometimes with modifications peculiar to the species (e.g., echolocation in bats). Not only are the anatomical pathways similar, but so are the qualitative responses of individual neurons, neural assemblies, and whole animals to the same stimuli and to pharmacological interventions. The argument then is: if the brains are so similar and if one

Table 4.1 Comparison of Neurons in the Nervous System and Brains of Various Species

Species	Total Neurons	Cortical Neurons
C. elegans	302	
Sea slug	18,000	
Fruit fly	250,000	
Ant	250,000	
Honey bee	960,000	
Frog	16,000,000	
Mouse	71,000,000	4,000,000
Rat	200,000,000	18,000,000
Pigeon	310,000,000	
Octopus	500,000,000	
Dog		160,000,000
Cat	760,000,000	300,000,000
Raven	2,171,000,000	
Rhesus monkey	6,376,000,000	480,000,000
Chimpanzee		6,200,000,000
Human	86,000,000,000	21,000,000,000
African elephant	257,000,000,000	11,000,000,000
Dolphin		37,200,000,000

Source. https://ipfs.io/ipfw/ . . . /wiki/List_ofanimals_by_number_of_neurons.html)

species (*h. sapiens*) is known to have consciousness, is it not likely that this feature will be a property of the other brains also? This was, of course, precisely the argument developed by Thomas Huxley.

The neuronal population results for the octopus are of special interest, in that they take the basis for consciousness further. This species, which, as already noted, is so capable of intelligent behavior, has approximately half a billion neurons distributed in a nervous system totally dissimilar to those of mammals. In Figure 4.3 a central neuropil is shown surrounding the esophagus and flanked on each side by a head ganglion and prominent optic lobe. However, a large proportion of the octopus neurons are contained in complex neural structures innervating the sensory surfaces and muscles of the suckers lining the 8 tentacles. Indeed, it appears as if the octopus has a number of distributed brains, and, consistent with this view, a tentacle that has been severed from the body of the creature remains capable of self-directed movement. Most importantly, the octopus not only suggests that consciousness can be present in the absence of a cerebral cortex but, since the octopus is an invertebrate, that consciousness can appear in an evolutionary line far removed from that leading to mammals.

The Cambridge Declaration on Consciousness

In the summer of 2012 a number of eminent neuroscientists came together in Churchill College, in the University of Cambridge in the United Kingdom, to take part in the Francis Crick Memorial Conference on Consciousness in Human and Non-Human Animals. Toward the end of the meeting a small group drew up a statement that was then signed by all the participants at a dinner held the same evening. And so, on July 7, 2012, in the presence of Stephen Hawking, the world's leading astrophysicist, the following declaration was proclaimed:

> We declare the following: "The absence of a neocortex does not appear to preclude an organism from experiencing affective states. Convergent evidence indicates that non-human animals have the neuroanatomical, neurochemical, and neurophysiological substrates of conscious states along with the capacity to exhibit intentional behaviors. Consequently, the weight of evidence indicates that humans are not unique in possessing the neurological substrates that generate consciousness. Non-human animals, including all mammals and birds, and many other creatures, including octopuses, also possess these neurological substrates."

The Declaration had appeared 141 years after the publication of Darwin's *Descent of Man* and 36 years after Griffin's *The Question of Animal Awareness*. While appreciating the importance of the Declaration, a thoughtful observer could only wonder why it should have taken so long for the neuroscientists to have made it.

Notes and References

1. No better example of the disregard for animal life exists than the 19th-century slaughter of the plains buffalo (bison) in the central United States. Within a few years an estimated population of 25 million buffalo had been reduced to fewer than a hundred animals. So little thought was given to the massacre that trains would stop when a herd was sighted, giving passengers the opportunity to get out their firearms and blaze away at the magnificent but unsuspecting creatures. In the end, the only reminders of the animals were the huge heaps of bones littering the desolate grasslands.
2. Darwin C. *The descent of man and selection in relation to sex*. London, England: John Murray, 1871.
3. Darwin C. *Journal of researches into the geology and natural history of the various countries visited by H.M.S. Beagle*. London, England: Henry Colburn, 1839. (Subsequent editions were simply titled "The Voyage of the Beagle").

4. Darwin C. *On the origin of species by means of natural selection, or the preservation of favoured races in the struggle for life*. London, England: John Murray, 1859.
5. Darwin's struggle to identify the evolution of consciousness is the subject of a fascinating article by the late Professor Chris Smith: Smith CU. Darwin's unsolved problem: the place of consciousness in an evolutionary world. *Journal of the History of the Neurosciences*. 2010; 19(2): 105–120.
6. Clifford WK. On the nature of things-in-themselves. *Mind*. 1878; 3: 57–67.
7. See the suggestion by David Chalmers in Chapter 14.
8. Huxley TH. On the hypothesis that animals are automata, and its history. *Nature*. 1974; 10: 362–366. (Clark University has published the lecture on the Internet: aleph0.clarku.edu/Huxley/CE1/AnAuto.html).
9. Griffin DR, Galambos R. The sensory basis of obstacle avoidance by flying bats. *Journal of Experimental Zoology*. 1941; 86: 481–506.
10. Griffin DR. *The question of animal awareness*. New York, NY: Rockefeller University Press, 1976.
11. Griffin DR. *Animal thinking*. Cambridge, MA: Harvard University Press, 1984.
12. Griffin DR. *Animal minds*. Chicago, IL: University of Chicago Press, 1992.
13. Typing "Animal Intelligence" into Google can become the beginning of a very interesting journey. The cited examples of crow intelligence were gleaned from the Internet, as was the remarkable ability of the collie, Chaser.
14. Pepperberg IM. Cognition in the African Grey parrot: preliminary evidence for auditory/vocal comprehension of the class concept. *Animal Learning & Behavior*. 1983; 11: 179–185.
15. A recently published book devoted to consciousness in the octopus is Godfrey-Smith P. *Other minds: the octopus, the sea, and the deep origins of consciousness*. New York, NY: Farrar, Strauss & Giroux, 2016.
16. Scientific notes. *Appleton's Journal*. 1873; 10(July 19): 94.
17. Jerome Lettvin in interview with Dr. Lincoln Stoller (ls@tengerresearch.com). Reprinted by courtesy of Dr. Stoller.
18. Fulton JF. Sir Charles Scott Sherringon (1857–1952). *Journal of Neurophysiology*. 1952; 15: 167–190.
19. Gardner RA, Gardner BT. Teaching sign language to a chimpanzee. *Science*. 1969; 165: 664–672.
20. Inoue S, Matsuzawa T. Working memory of numerals in chimpanzees. *Current Biology*. 2007; 17: R1004–R1005.
21. Silberberg A, Kearns D. Memory for the order of briefly presented numerals in humans as a function of practice. *Animal Cognition*. 2009; 12: 405–407.
22. Ryabov VA. The study of acoustic signals and the supposed spoken language of the dolphin. *St Petersburg Polytechnical University Journal: Physics & Mathematics*. 2016; 2: 231–239. Also: http://dx.doi.org/10.1016/j.spjpm.2016.08.004
23. McMahan EA. Bugs angle for termites. *Natural History*. 1983; 92: 40–47.
24. Nagel T. What is it like to be a bat? *The Philosophical Review*. 1974; 83(4): 435–450.
25. Perhaps this is too negative. We can, for instance, get a glimmer of what it is like to be a bat from humans who, blind from birth, have learned to navigate using echoes from clicking or tapping.

5

Groundwork

Cellular Events

Before trying to understand how different parts of the brain combine in the phenomenon of consciousness, several basic pieces of information must be considered. These include the structure of the nerve cell ("neuron") and the nature of the connections it makes with other neurons. Various roles of the glial cells are also important. Lastly there is the nature of the nerve impulse and the excitation and inhibition it is able to bring about in other nerve cells.

The Nerve Cell

Anatomy and physiology . . . form and function. How better to approach a biological problem than to first see what clues might emerge from the study of structure? This was the approach used by Santiago Ramón y Cajal (1852–1936; Figure 5.1),[1] considered by many to have been the greatest neuroscientist of all. That the Spaniard should have won such respect, let alone become a scientist, borders on the miraculous.

Cajal was born in Petilla de Aragón, a small community in the foothills of the Pyrenees. His father was a doctor who made a precarious living performing minor procedures while travelling from one poverty-stricken village to another. A rebellious youth, Cajal's main interests had been in art and the observation of nature. Eventually acquiescing to his father's demand to follow him in his profession, he began his anatomical studies on bones stolen from the local cemetery. His interest piqued, Cajal enrolled in the medical school in Zaragoza. Following graduation he had a military posting to Cuba, which was then in the middle of a colonial rebellion. After being seriously ill with malaria and then with tuberculosis, Cajal recovered and returned to Zaragoza. There he bought a second-hand microscope, read the research papers coming from France and Germany, and began a serious study of anatomy. The hard work paid off. Appointed to a chair in anatomy at the university in Valencia, Cajal was given a superior light microscope, a German Zeiss, to complement the Reichert microtome he was using to cut tissue sections. Then came the chance observation that would determine his life's future course. On a visit to a laboratory in Madrid he was shown slides

Figure 5.1. Santiago Ramón y Cajal posing with a Carl Zeiss microscope. Photo by Zeiss Microscopy from Germany (Wikimedia Commons CC BY-SA 2.0)

of brain tissue that had been stained by a new method, one that had been developed by his contemporary, the Italian anatomist Camillo Golgi (1843–1926; Figure 5.2).[2]

Cajal was entranced by the slides. For the first time he saw individual neurons with all their branches, and to him they were beautiful—so much so that he decided to devote the remainder of his life to studying them. Soon after embarking on this new journey, he began experimenting with Golgi's staining method. By adding osmium tetroxide to the fixative, he found that he could get better penetration of silver chromate into the cell bodies and processes of the neurons. Cajal, like others after him, must have wondered why it was that the Golgi stain only picked out a very small proportion of the neurons in the specimen. What was it that had made these cells different from the others? It was an obvious and interesting question, and one that has yet to be answered.[3]

In his enthusiasm Cajal studied one part of the brain after another, moving between species as he did so and often preferring immature brains because of their lesser complexity. Retina, cerebellum, brain stem, cortex—all of these were

Figure 5.2. Camillo Golgi. Though they shared a Nobel Prize, both Golgi and Cajal were adamant in their disagreements over the fine structure of the nervous system.

examined. To record what he saw, Cajal repeatedly adjusted the plane of focus of his microscope so that he could make out the full structure of a particular neuron, which he then drew. Befitting a man who in his boyhood had considered art as a profession, the detailed pen-and-ink sketches were exquisite (Figure 5.3).

There was a problem, however. Cajal realized that, no matter how excellent his publications, they were not attracting attention. In a field dominated by Germans, but with major contributions from other central European countries, who was going to pay attention to an obscure Spaniard? There was only one thing to be done—he would attend a major scientific meeting and force recognition of his work. And so, in 1889, he placed himself in Berlin at a meeting of the German Anatomical Society. He took his precious Zeiss microscope with him, and, like many of the other attendees, set up a demonstration in one of the large rooms.

Preoccupied with their own demonstrations, none of the other participants showed much interest in Cajal's work, however. Frustrated, Cajal then buttonholed one of the leading neuroanatomists, the Bavarian Albrecht Kolliker, and persuaded him to look through the Zeiss and to see the sketches that Cajal had

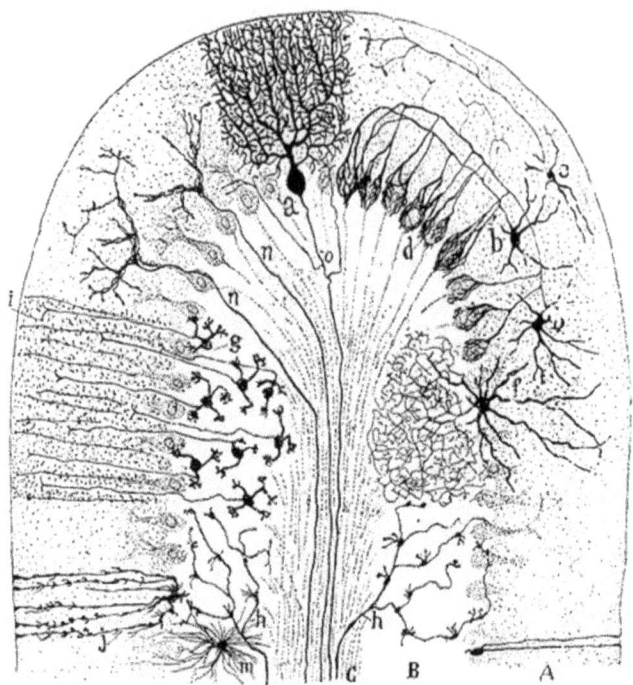

Figure 5.3. Cajal's beautiful drawing of the cell types found in the cerebellum.

brought with him. Cajal most likely talked to the older man about his novel findings and their significance. It was the turning point—Kolliker, impressed and generous, gave Cajal the support and recognition he craved for his work and his ideas.[4] But what were those ideas?

Cajal's main proposition was that the brain and spinal cord were composed of nerve cells that were separate from each other. This was in contrast to the prevailing belief that there were protoplasmic bridges that linked the elements of the nervous system together in one vast net. One of the main proponents of this "reticular" hypothesis was the inventor of the histological staining method that Cajal was using, Camillo Golgi. And, ironically, one of the strongest arguments against the reticular concept came from the Golgi method—how could the stain be so selective unless the neurons were physically separate from each other?

Another important concept of Cajal's concerned the "polarization" of the nerve cell, that is, the direction in which the nerve impulses normally traveled. Again, this was a controversial topic, but, having studied many types of neuron in a variety of sites and species, Cajal reasoned that it must be from the dendrites to the axons. Through their many branches the dendrites collected information and passed it on via the cell's axon. This concept was extended to sensory neurons in the periphery, where the receptor endings embedded in skin and other tissues were seen as the equivalent of dendrites. It is interesting that, in relation to the dendrites, Cajal's technique and sharp eyes had been able to examine the minute projections from the dendritc shafts, the dendritic "spines" (Figure 5.4).[5] Although such irregularities had been observed by others, they had been attributed to staining artifact. As the electron microscope would show many years later, and as Cajal argued, the dendritic spines were specialized to receive information from the endings of fibers (axons) belonging to other nerve cells. These meeting points were later termed "synapses" by Charles Sherrington (see later discussion).

Cajal's ideas, important as they were, were superimposed on a vast study of the architecture of the nervous system. Moreover, Cajal was interested in how the nervous system developed and in how nerve fibers recovered from injury. Though it seemed hardly possible, given the limited resolution of the light microscope, Cajal described the "growth cones" at the ends of regenerating fibers. Trusting his intuition as to how structures in the nervous system worked, Cajal believed that the growth cones were sensing some signal in the denervated tissue.

This last speculation—that there were chemicals in the tissue that guided the growth of nerve fibers—was one that was taken up many years later and

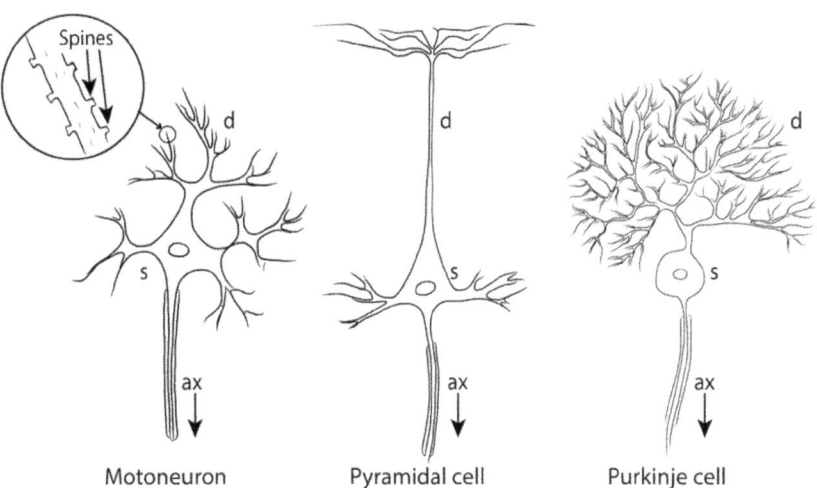

Figure 5.4. Three types of neuron with characteristic morphologies; *ax*, axon; *d*, dendrite; *s*, soma (cell body). The pyramidal and Purkinje neurons exist in the cerebral and cerebellar cortices respectively, while the motoneurons are in the brain stem and spinal cord.

approached from two entirely different directions, one of them involving Victor Hamburger (1900–2001).[6] Hamburger was born near Breslau in what was then eastern Germany. The son of a prosperous textile manufacturer, he became interested in natural history at an early age and followed his inclination while attending university. While a doctoral student in Hans Spemann's department in Freiburg, he witnessed the dramatic results of grafting tissue, containing a hypothetical "organizer," on to an amphibian embryo so as to produce an attached second embryo—work for which Spemann would win the Nobel Prize. It was some years later, while on a Rockefeller Fellowship in the United States, that Hamburger learned that he had lost his university position in Germany as part of the Nazi "cleansing" process. Remaining in the United States, Hamburger accepted a position in 1935 at Washington University, St. Louis, where, in the zoology department, he would remain for the rest of his long academic life.

Using the chick embryo as his preferred experimental model, Hamburger made several important discoveries,[7] one of the earliest being that there was a quantitative relationship between the amount of tissue and its innervation. Excision of a limb bud resulted in the presence of fewer motoneurons in the spinal cord, while grafting an extra limb bud had the opposite effect—a higher than normal number. Evidently some kind of stimulus was passing in the nerve fibers from the periphery to the developing spinal cord that was necessary for the survival of the motoneurons. The first such nerve "trophic" factor to be found followed experiments in mice that had had tumor cells implanted abdominally; however, this factor—subsequently termed nerve growth factor (NGF)—exerted its effect on sympathetic and sensory neurons rather than on motoneurons. But the start had been made and, over the years, other trophic factors would be discovered. It was fitting that a Nobel prize should have been awarded in 1986 for the discovery of NGF and that one of the recipients was Rita Levi-Montalcini (Box 5.1), who had left Italy to work with Hamburger. The other recipient was Stanley Cohen, responsible for purifying NGF and determining its chemical structure and also for discovering an epidermal growth factor. But what about Victor Hamburger, the person who had started the whole enterprise? Not for the first time the Nobel Committee came up short, and it was not surprising that there was consternation in the scientific community.[8]

The second line of evidence for the guidance of growing nerve fibers by chemical signals came from the work of Roger Sperry (1913–1994; Box 5.2). Sperry, an American, grew up in Connecticut and obtained a PhD in experimental psychology under Paul Weiss (1898–1989) in Chicago. Later, in Karl Lashley's

Box 5.1 Rita Levi-Montalcini (1909–2012)

Before Rita Levi-Montalcini died she had been, at 103, the oldest living Nobel Laureate.

She was born into a wealthy Jewish family in Turin, Italy, where her father was an electrical engineer and mathematician and her mother an artist. Having initially contemplated a career as a writer, Rita Levi-Montalcini chose medicine instead, enrolling in the University of Turin Medical School. After graduating MD in 1936 she decided to specialize in neurology and psychiatry, having been inspired by the teaching of Giuseppe Levi, the noted neurohistologist. Her academic aspirations were abruptly compromised, however, by the edict of the dictator Benito Mussolini that Jews be denied careers in the professions. Undaunted, Levi-Montalcini's bedroom became her laboratory, and it was there that she undertook research on the developing nervous system, with Giuseppe Levi joining her during the early part of World War II. Her research project had been prompted by a paper by Victor Hamburger in St. Louis, Missouri, in which he had shown that removal of a limb bud in a chick embryo resulted in fewer motoneurons in the spinal cord (see main text). When Turin became unsafe because of bombing, her family moved to the country and then, under pressure from the Germans, to Florence. Regardless of the stage of the war, Levi-Montalcini kept at her research, and when the war was finally over she was able to resume her position in the University of Turin.

In 1947 she accepted an invitation from Hamburger to visit St. Louis and to repeat some of her work. She was able to convince Hamburger that the reduction in the number of motoneurons following limb bud extirpation was due to degeneration of existing neurons rather than to reduced production of new ones. Invited to stay at Washington University, Levi-Montalcini did so, becoming full professor in 1958. By then she was pursuing the ability of sarcoma cells (and later snake venom and mouse salivary glands) to promote growth of sensory and sympathetic neurons. She used tissue cultures of neurons for her assays of the active principle. This work led to the purification and chemical identification of nerve growth factor by her colleague, Stanley Cohen, and to Nobel Prizes for both of them in 1986. Eventually returning to Italy, Levi-Montalcini continued in research and became a prominent national figure, as well as a life Senator in the Italian parliament.

Box 5.2 Roger Sperry (1913–1994)

It is not often that a neuroscientist becomes famous for two entirely unrelated types of study, but this was the accomplishment of Roger Wolcott Sperry. Born in Hartford, Connecticut, Sperry was brought up by his mother, his father having died relatively young. Sperry was not only a good scholar at high school but a serious athlete, setting a state record for throwing the javelin. Moving to Oberlin College, Ohio, Sperry obtained a BA in English and then switched to science, gaining a master's degree in experimental psychology. Then came doctoral studies at the University of Chicago under Paul Weiss who, like Victor Hamburger in St. Louis, was a zoologist and a major figure in developmental neurobiology. Sperry chose to work on rats for his initial studies on neural development, crossing the nerves supplying two antagonist muscle groups in the hind leg. The prevailing theory in the laboratory was that nerve function would adjust to the use of the target organ, in this case muscle, but Sperry's results proved otherwise—the hindlimb movements remained awkward after the cross-innervation. It seemed, then, as if the connections within the nervous system were genetically programmed rather than dependent on use.

After Chicago, Sperry moved to Boston to work under Karl Lashley at Harvard. However, it was while he and Lashley were together at the Yerkes primate laboratory in Florida that Sperry had the idea of using the amphibian eye as a preparation to study the formation of neural connections (see main text). Not only did the results confirm the conclusions reached in the earlier studies on rats—that nerve connections were predestined and "hard-wired"—but it seemed possible to Sperry that developing or regenerating nerve fibers grew along paths determined by the presence of chemical signals in the tissues. The "chemoaffinity hypothesis" was subsequently validated in other laboratories by the discovery of nerve growth factor and other trophic molecules.

By 1952 Sperry had moved to the California Institute of Technology in Los Angeles where he had the opportunity to study patients who had had the corpus callosum divided as a last resort for the treatment of epilepsy. Against the prevailing wisdom that severing the bridge between the two hemispheres had little, if any, consequences for normal brain function, Sperry was able to show that the two sides of the brain differed markedly in their capacities. It was for this work, continued by his former graduate student, Michael Gazzinaga, that Sperry was awarded a Nobel Prize in 1981.

Source. Todman D. History of neuroscience: Roger Sperry (1913–1994). *IBRO History of Neuroscience*, 2008. http://www.ibro.info/Pub/Pub_Main_Display.asp?LC_Kocs_ID=3473.

laboratory in Florida, Sperry followed his interest in the development of neural pathways by some ingenious experiments on newts and then on adult frogs.[9] These involved cutting the optic nerves and allowing the fibers to regenerate to their normal destination, the optic tectum. By observing the attempts of the operated animals to capture flies placed within their visual fields, he was able to show that, rather than a haphazard return, the regenerating optic nerve fibers had found their way back to their original targets. The same directed growth occurred if, in addition to cutting the optic nerve, the eye was rotated so as to lie upside down; as a consequence the frog would attempt to catch a fly placed above it by striking downwards (Figure 5.5). From these and other types of experiments Sperry developed a chemoaffinity hypothesis that posited that a growing nerve fiber was directed to a specific position in the target tissue by sensing the concentration of a local chemical attractant. The hypothesis is now generally accepted as true, and NGF and other nerve trophic factors have been shown to be the guiding molecules in developing nerve pathways.

In the optic tectum, as in other parts of the nervous system, the point of functional contact between two nerve cells is known as a "synapse"; the term was introduced after considerable thought by Charles Sherrington. A synapse has four parts (Figure 5.6)—the terminal knob of an axon, the membrane of the dendritic spine or cell body of a second neuron, the narrow space ("cleft") between the two membranes, and a surrounding glial cell. Long after Sherrington's insight, the high magnification achieved with the electron microscope made

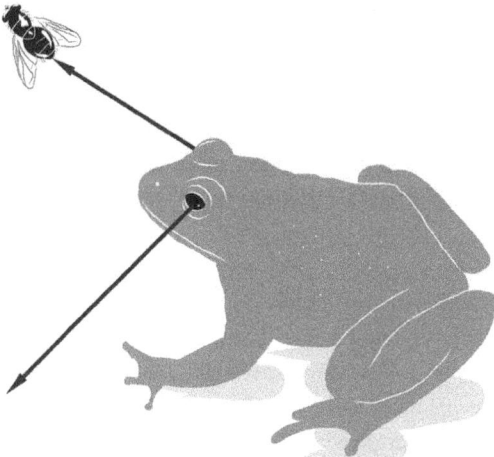

Figure 5.5. One of Sperry's experiments. The frog's remaining eye was rotated upside down, the optic nerve having been cut. Following nerve regeneration an object above the frog (here, a fly) would appear to be in the lower visual field.

58   SHERRINGTON'S LOOM

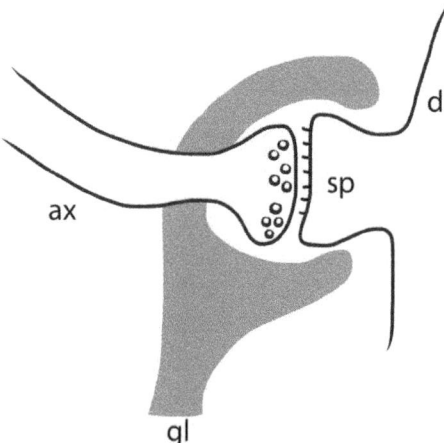

Figure 5.6. The components of a synapse. The knob at the end of the axon (*ax*) contains synaptic vesicles (rounded structures) containing neurotransmitter. Opposite the axon knob is one of the spines (*sp*) on the dendrite (*d*), with receptors protruding from the spine membrane. The axon and spine are separated by a narrow cleft and are enveloped by a glial cell (*gl*).

it possible to see small rounded objects in the axon terminal. Following their purification, these "synaptic vesicles" were shown to contain chemical neurotransmitters.

Over the years, and from a variety of sources, it became evident that a synaptic connection in the brain or spinal cord, once formed, was not necessarily fixed. Unless the synapse was repeatedly activated following the arrival of impulses in the axon terminal, it was liable to degenerate. Conversely, frequent usage of a synapse resulted in its strengthening and was likely to be instrumental in the forming of memories and in stabilizing other neural pathways. This concept of usage determining synaptic structure at a molecular level was attributed to Donald Hebb.[10]

Hebb (1904–1985; Figure 3.3) was born and raised in Atlantic Canada and tried school teaching as a career before deciding, just as Sperry did later, to study experimental psychology under Karl Lashley. Hebb followed Lashley from the University of Chicago to Harvard University, where, in 1936, he completed a PhD thesis on the effects of early visual deprivation in the rat. In the following year Hebb accepted a position with the prominent neurosurgeon Wilder Penfield at the Montreal Neurological Institute. Hebb was but one of several Montreal neuroscientists grouped around Penfield who would make major contributions to the field of consciousness. Hebb's deduction that usage was important for the maintenance of synaptic connections appeared in his influential book, *The*

Organization of Behavior, published in 1949, at a time when he was professor of psychology at McGill University.[11] As Hebb put it:

> When an axon of cell A is near enough to excite cell B and repeatedly or persistently takes part in firing it, some growth process or metabolic change takes place in one or both cells such that A's efficiency, as one of the cells firing B, is increased.

This proposition became known as "Hebbian learning." Nowadays, Donald Hebb's carefully constructed sentence is expressed more succinctly: "Neurons that fire together wire together."

Half a century later, the New York neuroscientist Eric Kandel would be awarded the Nobel prize for working out the molecular basis of Hebb's synaptic strengthening.[12]

Glial Cells

Cajal did not limit his exploration of the nervous system to neurons.[13] There were other cells that had been discovered by Rudolf Virchow and that the German anatomist had termed "neuroglia." Virchow envisaged these cells as part of a "cement" that held the neurons in place. Others were to give better descriptions of the glial cells, but it was Cajal who subjected them to the most comprehensive study and who thought most deeply about their possible functions. Cajal observed that the most prominent type of glial cell, the astrocyte, formed endfeet on the walls of blood vessels; this arrangement led him to suppose that the astroglia could control local blood flow to the neurons, depending on which parts of the brain or spinal cord might be particularly active. This function has been validated, though by the release of vasoactive peptides from the astrocytes rather than by mechanical contraction and relaxation of astrocytic processes in the way that Cajal envisaged. The astrocytes, by virtue of their intimacy with blood vessels and neurons, are also known to mediate the supply of nutrients and oxygen to the neurons, and to remove the waste products of the latter. Cajal also speculated that the astrocytes might control interactions between neurons by means of protoplasmic processes that could advance or retract between the nerve cells. Modern microscopic techniques have suggested that such a mechanism may, in fact, exist. Better known, however, is the role that astrocytes play in modulating synaptic function by removing transmitter from, or releasing it into, the synaptic cleft.

Yet another of Cajal's prescient ideas was that the "radial glia"—those cells extending outwards from the central canal of the cord and the ventricles of the brain and that are now known to guide migrating neurons—could themselves differentiate into neurons. Finally, he and his former student, Rio-Hortega, identified a different type of glial cell, the oligodendrocyte, that would later be shown to wrap layers of myelin around axons.

Though the neuroglia, like other cells in the body, have membrane potentials, they cannot fire impulses and are therefore unlikely to contribute directly to consciousness. Nevertheless, they have been shown to communicate with each other through gap junctions using calcium signals and—through the variety of means already described—to modulate synaptic activity between neurons and hence influence consciousness indirectly.

The Nerve Impulse

At the time that Cajal was working out the detailed structure of the nervous system, important advances had already been made into understanding the nature of the nerve impulse, the brief electrical signal that sped along the nerve fiber to end at its synaptic terminals.[14] It was the impulses that brought the brain to life or, in Sherrington's analogy: "In the great head-end which has been mostly darkness spring up myriads of twinkling stationary lights and myriads of trains of moving lights of many different directions."

Following Galvani's chance observation that there was such a thing as "animal electricity," Emil du Bois-Reymond (1818–1896), in Berlin, used his sensitive galvanometer to show that electrical stimulation caused a transient negativity in a peripheral nerve. Demonstrating that this negativity traveled along the nerve with a finite velocity was one of the achievements of the extraordinary polymath, Hermann Helmholtz (1821–1894; Box 10.2). It was left to Julius Bernstein (1839–1917; Figure 5.7)—yet another member of the German states—to show that a potential difference existed across the membranes of nerve and muscle fibers and that this was the result of the cell membranes being semipermeable.

The ingenious Bernstein did something else. He designed an apparatus that, in conjunction with a galvanometer, enabled the form of the traveling negativity (the nerve impulse) to be reconstructed from short time segments (Figure 5.8); this elegant approach overcame the problem of the slow responsiveness of the galvanometer. This was in 1868, at a time when German science, including physiology, was still in the ascendant and yet had already surpassed its competitors. Nevertheless, there was a small contribution from Britain. In the primitive working conditions that characterized the Physiological Laboratory at Cambridge, Keith Lucas and Edgar Adrian were able to demonstrate that the nerve impulse obeyed an "all-or-none" law. By this they meant that, following an adequate stimulus, the nerve impulse would attain its full size and propagate

Figure 5.7. Julius Bernstein (1839–1917). Bernstein was appointed head of the physiology department at the University of Halle at 34 and became rector of the university later. For many years he was the foremost authority on the ionic mechanisms underlying the resting and action potentials of nerve and muscle fibers.

down the fiber; if the stimulus was insufficient, the impulse would fail altogether. In comparison with the elegant studies of Bernstein, it was rough work, but the minds behind it were as brilliant as any. Though Lucas would be killed in a flying accident during World War I, his young protégé, Adrian, would go on to become the most successful neurophysiologist of his time.

The penultimate step in the understanding of the nerve impulse was the employment by Alan Hodgkin (1914–1998) and Andrew Huxley (1917–2012) of the squid giant axon for their experiments. With this preparation the two investigators (Figure 5.9) were able to show, by inserting a capillary glass recording electrode inside the fiber, that the impulse was associated with a brief reversal of the normal polarity across the membrane, just as Bernstein had shown indirectly with his differential rheotome. Further, the reversal and the subsequent resumption of the normal potential resulted from transient changes in membrane conductance, that is, in the permeability of the fiber membrane to

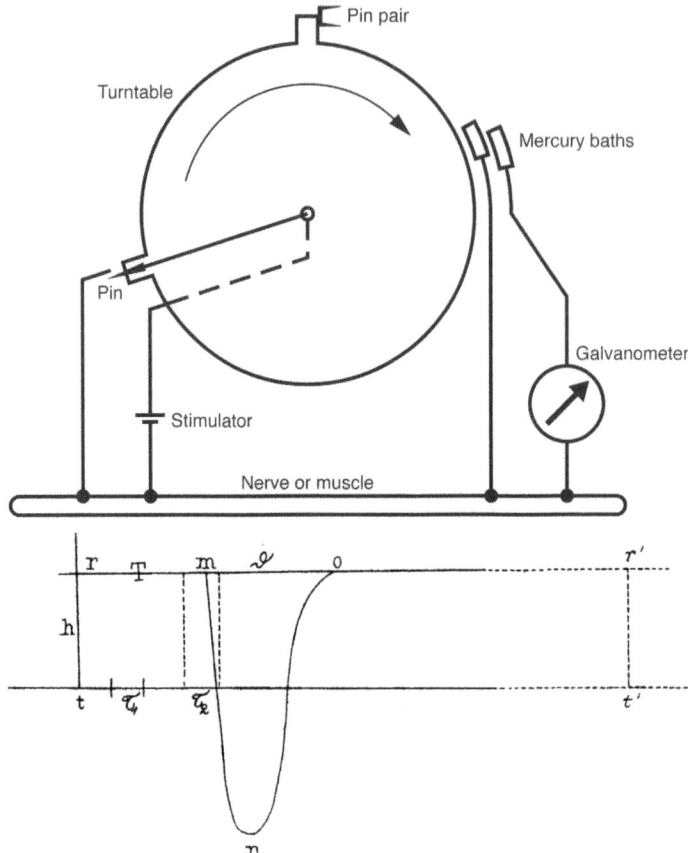

Figure 5.8. *Top.* Bernstein's differential rheotome for reconstructing the muscle or nerve fiber impulse (action potential). As the turntable rotated, a pin completed a circuit for stimulating the fiber. A moment later a pair of pins, dipping into two mercury baths, completed a recording circuit. By altering the relative positions of the two baths, different segments of the action potential could be recorded by the galvanometer. Because of the slow response of the latter, the turntable had to rotate multiple times for the full deflection to register. *Bottom.* Reconstruction of the full action membrane (*m,n,o*); the peak briefly exceeds the potential across the resting potential. (From Bernstein J. *Archiv fur die Gesamte Physiologie des Menschen und der Tiere*, 1868; 1: 173–207)

certain ions. There was a short-lived increase in sodium conductance followed by a slower increase in potassium conductance.

The final step in the impulse research was one considered technically impossible by those working in the field. It was to show, by X-ray diffraction, the conformational changes in the ion channels that produced the momentary alterations

Figure 5.9. Alan Hodgkin (on right) and Andrew Huxley inspecting something, possibly during their work together in Plymouth in 1939. (Courtesy of Professor Bill Harris of the Department of Physiology, Development and Neuroscience, University of Cambridge)

in membrane conductance. Despite the naysayers, and despite initially limited resources, Roderick Mackinnon (1956–), working at the Rockefeller University in New York, was able to achieve success—in part by studying the ion channels of bacteria as well as those of nerve cells. For this impressive achievement Mackinnon, like Hodgkin and Huxley before him, would be awarded the Nobel Prize in Physiology or Medicine.

Excitation and Inhibition at the Synapse

The nerve impulse, having traveled down the nerve fiber, then affects the neurons that form synapses with the multiple endings of the fiber. Whether there is excitation or inhibition depends on the identity of the chemical transmitter released by the fiber endings into the narrow space separating the membranes of the respective neurons.

The credit for the first systematic studies of excitation and inhibition in the nervous system is usually given to the eminent British physiologist Charles Sherrington, but this is mistaken. It is generally not realized that, in the decades leading up to the 1917 Revolution, there was a flowering of the sciences—biological, physical, and chemical—in tsarist Russia. That country's contributions were reflected in the choice of Moscow for the International Medical Congress in 1897. In 1904 Ivan Pavlov won a Nobel Prize in Physiology or Medicine for his work on conditioned reflexes and, four years later, Ilya Mechnikov achieved the same distinction, in this case for immunology. But before Pavlov and Mechnikov, and the 1897 Congress, there was the work of Ivan Sechenov (Figure 5.10).[15]

Sechenov was born in 1829 in a village some 400 kilometers east of Moscow. As the son of a landowner in a remote area, Sechenov was privately schooled prior to entering the military engineering school in St. Petersburg at the age of 14. After graduating and serving two years in the army, Sechenov had enrolled in the faculty of medicine at the University of Moscow. Following his degree in 1856, he had spent three years among the leading university laboratories in Germany and Austria acquiring knowledge and skills in research. Within two years of his return to Russia in 1860, Sechenov was off again, this time to spend a year in the laboratory of Claude Bernard in Paris. It proved to be a very rewarding year for it was then, in experiments on frogs, that he discovered that spinal reflexes could be inhibited by stimulation of the brain. Sechenov gave the first description of this "central inhibition" in a paper published in 1863. Sechenov was already

Figure 5.10. Ivan Sechenov. (Wikimedia Commons)

familiar with the techniques and apparatus used to investigate peripheral nerve and muscle, having spent time with Emil du Bois-Reymond in Berlin, and he was thus able to establish his own electrophysiological laboratory and to continue his studies of the nervous system. In relation to consciousness, his most notable proposition was that human behavior was directed by reflexes, a startling idea that was taken up in earnest by his younger countryman, Pavlov. With ample reason, Sechenov came to be regarded as the "father" of Russian physiology.

There has been speculation that, during the 1897 International Medical Congress in Moscow, Sechenov, as president of the Section on Physiology and Physiological Chemistry, may have met Richard Caton, a delegate from Britain who was the first to record ongoing electrical activity in the brain.[16] Whether true or not, it was in Britain that the next advances in excitation and inhibition would take place.

Although Charles Scott Sherrington[17] had not attended the 1897 Congress, he was aware of Sechenov's work and had been present at a similar meeting held in London 16 years previously. At the time he had been 24 and still a medical student. However, it was that 1881 Congress that would shape his career, for it was then that he witnessed the presentation by David Ferrier,[18] one in which he showed that removal of part of a monkey's brain resulted in paralysis on the opposite side of the body. For Sherrington, Ferrier's demonstration was his introduction to experimental neurology. Like Sechenov, Sherrington would combine his studies of the nervous system with those of other organs and tissues. He became a skilled anatomist and neuropathologist and would lecture on histology. He was interested in the composition of blood, including its changes during infections. He studies cholera, diphtheria, and tetanus. Vision was another subject that interested him, especially flicker fusion. However, it was his investigations of the nervous system that brought him fame.

Sherrington demonstrated the presence of receptors in muscles, tendons, and joints that were responsible for proprioception, the body's awareness of itself. In addition, the muscle spindles were shown to provide the stimulus for stretch reflexes. Sherrington, who was a fine dissector, was able to divide selected dorsal nerve roots in animals and thereby show the area of skin ("dermatome") supplied by each root. Later experiments, discussed in the next chapter, undertaken while he was professor of physiology in Liverpool, enabled him to map out the representation of different movements in the motor cortex of various primates. But his greatest work, that involving excitation and inhibition in the nervous system, came from his many investigations of spinal reflexes. This work had been started in Liverpool but reached its pinnacle after his appointment as Waynflete Professor of Physiology at Oxford. Although the dissections of spinal cord, nerve, and muscle required skill, the recording apparatus was simple—a muscle tendon was attached to a lever that wrote a trace on smoked paper wrapped around a rotating drum. This apparatus, the "kymograph," had been the invention of Carl

Ludwig in Germany many years before, and it was only in his later years that Sherrington combined the recordings of tension with those of electrical activity in the muscles. Nevertheless, the experiments were good enough to demonstrate a number of key features of reflex activity. For example, the fact that two weak stimuli could produce a greater effect than the sum of the responses to separate stimuli was evidence for subliminal excitatory fringes surrounding the discharging motoneurons in the spinal cord. And a stimulus that produced a "central excitatory state" in one group of motoneurons was likely to cause a "central inhibitory state" in motoneurons supplying antagonist muscles.[19,20]

But what were these excitatory and inhibitory "states"? To answer this question required a very different approach, one involving the insertion of a glass microelectrode tip through the membrane of a motoneuron in the spinal cord. Sherrington retired long before this technology became possible, but he lived long enough to see the results achieved by his former doctoral student, John Eccles.[21] Eccles showed that an excitatory stimulus caused a transient reduction in the potential across the membrane of the motoneuron (an excitatory postsynaptic potential). Conversely, an inhibitory stimulus resulted in a brief increase of the membrane potential (the inhibitory postsynaptic potential).[22] In the former case a neurotransmitter, released into the synaptic cleft, opened sodium channels in the recipient motoneuron; in the latter case, a different transmitter opened chloride channels. Though Sherrington's work, and that of Eccles, had been carried out on the spinal cord, it was realized that the same processes would probably apply to synapses elsewhere in the nervous system, and so it proved. Both Sherrington and Eccles would go on to win Nobel Prizes—Sherrington's in 1932 and Eccles' in 1963.

Regarding consciousness, Sherrington contemplated its relationship to the working of the brain throughout his career though, unlike Sechenov, he hesitated to put his ideas to paper. He was certainly no dualist, however, for it was "busy common sense ... to conjoin the mental with the physiological."

Numbers of Neurons and Glia

This account has not included any mention of numbers. How many neurons and glial cells are there in an "average" human brain, and how might one find this out? Obviously, it would be a life's work to count all the neurons in section after section of brain, and the Golgi stain, were it to be used, would only pick out a small percentage of the cells. In the mouse this problem was solved many years ago by Johnson and Erner.[23] They made a coarse homogenate of the mouse brain, stained the neurons with dye, diluted the homogenate, and then, using a light microscope and a blood cell counting chamber, determined the number of the neurons present in a known small volume. From the dilutions and the initial volume of the homogenate, it was easy to calculate the total number of neurons in the whole brain. The same principle, but using different techniques and many more samples, was employed decades later by Suzana Herculano-Houzel and her Brazilian colleagues for human brains.[24] The answer—that are were 86 billion neurons in the human brain—is one that is now widely cited, but it was only a rough estimate, since all 4 brains examined were male and the estimates ranged from 79 billion to 95 billion. Further, although the cerebral hemispheres make up approximately 82% of the brain's mass and were found to contain 72%of the glial cells by the Brazilian group, they only had 19%of the brain's neurons, some 16 billion of them; most of the brain's neurons were to be found in the cerebellum (69 billion).

Perspective

There is much more that could be said about neurons, impulses, and synapses in light of more recent findings. For example, synapses are now known to be present not only between axon and cell body or axon and dendrite but between axon and axon and dendrite and dendrite. And, in addition to the main potassium, sodium, and chloride conductances, there are others—even in the same neuron—that influence membrane excitability. Even Sir Henry Dale's old dogma that a neuron releases the same type of neurotransmitter at all its synapses is no longer valid, since recent studies have shown that more than one transmitter may be present in the same neuron, even in the same vesicle. But for a background to understanding conscious mechanisms, knowledge of such details is unnecessary. Rather, it is time to consider the next question:

> Which part of the brain is responsible for what function?

Notes and References

1. Cajal told the story of his extraordinary life in *Recuerdos de mi vida* (Recollections of my life), published in 1901. In 1989 the MIT Press brought out an English translation by E. H. Craigie in paperback.
2. A brief account of Golgi's life, including his major dispute with Cajal over the neuron doctrine, is given in Wikipedia and in McComas AJ. *Galvani's spark: the story of the nerve impulse*. New York, NY: Oxford University Press, 2011. A fuller account of the academic rivalry between the two neuroanatomists is in Rapport R. *The discovery of the synapse*. New York, NY: W. W. Norton, 2005.
3. See Koyama Y. The unending fascination with the Golgi method. *OA Anatomy*. 2013 Sep 1; 1(3): 24.

4. The remainder of Cajal's life is described in Reference 1. In recognition of his stature as a neuroscientist, the Spanish government created an institute for Cajal in Madrid. There he acted as mentor and inspiration for a number of younger neuroanatomists, including Lorente de Nó. Cajal died during the time of the Spanish Civil War.
5. Yuste R. The discovery of dendritic spines by Cajal. *Frontiers in Neuroanatomy*. April 21, 2015. https://doi.org/10.3389/fnana.2015.00018
6. Hamburger V. *The history of neuroscience in autobiography*. Squire L, ed. vol. 1. Washington, DC: Society for Neuroscience, 1958:222–250.
7. Another of Hamburger's findings was "programmed cell death." This refers to the natural loss of neurons during embryogenesis, first noted for motoneurons in the chick spinal cord.
8. Purves D, Sanes JR. The 1986 Nobel Prize in physiology or medicine. *Trends in Neurosciences*. 1987; 10: 231–235. Another notable exclusion from a Nobel Prize was George Bishop, whom many thought should have shared the 1944 award with Joseph Erlanger and Herbert Gasser (see McComas AJ. *Galvani's spark: the story of the nerve impulse*. New York, NY: Oxford University Press, 2011.
9. Perhaps the best paper for this type of experiment is Sperry RW. Optic nerve regeneration with return of vision in anurans. *Journal of Neurophysiology*. 1944; 7: 57–69.
10. Donald O. Hebb-Wikipedia. https://en.wikipedia.org/wiki/Donald_O._Hebb.
11. Hebb DO. *The organization of behavior: a neuropsychological theory*. New York, NY: Wiley.
12. A full account of Kandel' research, including his early work on the hippocampus, is given in his autobiography, written after his award of the Nobel Prize in 2000 (Kandel ER. *In search of memory: the emergence of a new science of mind*. New York, NY: W. W. Norton, 2007.
13. Garcia-Marin V, Garcia-López P, Freire M. Cajal's contributions to glia research. *Trends in Neurosciences*. 2007; 30(9): 479–487.
14. The history of research into the nerve impulse is given in McComas AJ. *Galvani's spark: the story of the nerve impulse*. New York, NY: Oxford University Press, 2011.
15. Sechenov IM. *International encyclopedia of the social sciences*. Encyclopedia. com. http://www.encyclopedia.com/social-sciences/applied-and-social-sciences-magazines/sechenov-ivan-m.
16. See Chapter 13.
17. See vignette at the beginning of this book.
18. See Chapter 6.
19. Eccles JC, Sherrington CS. Reflex summation in the ipsilateral spinal flexion reflex. *Journal of Physiology*. 1930; 69: 1–28.
20. Eccles JC, Sherrington C. Studies on the flexor reflex: VI-inhibition. *Proceedings of the Royal Society of London B*. 1932; 109: 91–113.
21. See Box 3.3.
22. The most readable and comprehensive account of Eccles' recordings of postsynaptic potentials in motoneurons is given in his monograph: Eccles JC. *The physiology of nerve cells*. London, England: Oxford University Press, 1957.
23. Johnson HA, Erner S. Neurone survival in the ageing mouse. *Experimental Ferontology*. 1972; 7: 111–117.
24. Azevedo FAC, Carvalho LRB, Grinberg LT, et al. Equal numbers of neuronal and nonneuronal cells make the human brain an isometrically scaled-up primate brain. *Journal of Comparative Neurology*. 2009; 513: 532–541.

6
Mapping the Brain

It is natural to inquire what things *do*. So far as the cerebral cortex is concerned, the question would be: Do all parts contribute to consciousness, or is there a special region that brings everything together—thoughts, emotions, sensations, and intentions? It was the possibility of such a conscious-creating integrator that led to Descartes' suggestion of the pineal gland, Penfield's of the brainstem, and, in present times, Crick and Koch's of the claustrum.[1] Alternatively if the entire cortex is involved in consciousness, as the contemporary philosopher Daniel Dennett has proposed,[2] what is the contribution of each region? It was a question that, centuries earlier, had intrigued Thomas Willis, Descartes' contemporary. With reference to the convoluted surface of the cerebral hemispheres, he wrote: "the animal Spirits, for the various acts of Imagination and Memory ought to be moved within certain and distinct limited or bounded places."

Remarkably, even by the middle of the 19th century, there were no answers to the question of localization of function in the cortex. This was at a time when the Industrial Revolution was well underway; when Michael Faraday had demonstrated the relationship between magnetism, electricity, and force; and the world—with the exception of the Polar regions—had been effectively mapped. But for the human brain, the generator of all this industrial and scientific activity, there was no map. True, the external features of the brain had been described long ago by Galen (130–210) and then, in increasing detail during and after the Renaissance, by Leonardo da Vinci (1452–1519), Vesalius (1514–1564), and Thomas Willis (1621–1675); moreover, the main structures had been given important-sounding Latin names—cortex cerebri, cerebellum, pons variolii, medulla oblongata, and so on. But as to function, there was almost nothing; indeed, for those who had been considering the possible functions of the cerebral cortex, there was one extraordinary happening that might have caused them to wonder whether the cortex had anything to do with consciousness at all.

In 1848 Phineas Gage was a successful 25-year-old American railway engineer, engaged in demolishing rock for a new rail track in Vermont. On this particular day, however, he was the victim of a blasting accident in which his own iron tamping rod, more than 3 feet long and slightly more than an inch in diameter, was driven right through his head, passing behind the left eye and taking a large part of the left frontal lobe with it. Astonishingly, Gage was able to speak and walk within minutes, even greeting his physician with the memorable words: "Doctor, here is business enough for you." Though Gage's life was cut short by the development of severe epileptic seizures, there is evidence that, for the first few years following the accident, there was little, if any, effect on his personality and, apart from immediate loss of vision from the injury to the left eye, none at all on his sensory or motor functions.

It was not until 1861 that the first clear evidence of cortical localization appeared, and it came from Paris. Pierre Paul Broca (1824–1880; Figure 6.1) had graduated early from medical school in that city, where he was later to hold professorial positions in anatomy, pathology, and surgery. A keen supporter of Darwin's evolutionary theory, Broca was also interested in anthropology and hypnotism and may have been the first to use hypnosis for surgical anesthesia. A further interest, that of brain function, led him to investigate a number of patients, all of whom had difficulty in speaking though not in understanding speech. At postmortem, the one region of the brain always found to be damaged was the third frontal convolution (the one immediately above the lateral sulcus; see Figure 3.2)—a site that has come to be known as Broca's area and as the center for the production of speech.

Figure 6.1. Portrait of Pierre-Paul Broca. (Wellcome Collection CC BY)

Motor Cortex

After the anatomical finding by Broca in Paris came an electrophysiological one from Berlin. Eduard Hitzig (1838–1907; Figure 6.2) was a young psychiatrist

Figure 6.2. Portrait of Eduard Hitzig and Gustav Fritsch, c. 1866. (Wellcome Collection) In the Wellcome catalogue the seated figure (6) is identified as Hitzig but is almost certainly Fritsch, with Hitzig the person (7) standing behind him. The presence of the dog suggests that the two were already experimenting, though their publication did not appear until four years later. With the possible exception of the helmeted figure, the others in the photograph are most likely faculty colleagues in the University of Berlin.

who wished to study brain function by means of electrical stimulation, and in this he was aided by Gustav Fritsch (1838–1927; Figure 6.2), an anatomist with wide scientific interests. Hitzig had previously attempted brain stimulation on wounded Prussian soldiers, and before that others had experimented on animals, mostly by excising or otherwise damaging parts of the cerebrum. In the latter line of investigation Hitzig was following others, the most prominent being the skillful French scientist Jean-Pierre Flourens (1794–1867), who had studied the effects of removing parts of the pigeon brain on behavior. Flourens had observed that, after taking away the pallium—the avian equivalent of the cerebral hemispheres—the birds were still able to fly when tossed into the air and they could swallow objects placed in their mouths, but they did not move spontaneously. He concluded from these experiments and from others on rabbits that the cerebral hemispheres were responsible for volition and sensation but not for the immediate production of movements. It was a view that was shared by the great majority of those interested in brain function.[3]

Such, then, was the situation at the time that Hitzig and Fritsch began studies of their own. Since the University of Berlin was unable to accommodate the planned experiments, the latter had to be conducted at home—allegedly on Frau Hitzig's dressing table. The initial experiments could only be described as brutal, with unanesthetized dogs crying with pain as they underwent partial removal of the skull so as to expose the underlying cortex. Later experiments did involve narcotization, however, and considerable care was taken over other aspects of the study, especially the localization of the observed effects. The two investigators found that, on applying weak currents through platinum wires attached to a battery, they could identify a region toward the front of the brain that elicited muscle contractions on the opposite side of the body. With experience, they could distinguish a map in which there were separate cortical sites for movement (or muscle twitching) of the leg, neck, and face. In two dogs, removal of the stimulated forelimb area produced weakness of the contralateral leg. Fritsch and Hitzig's results were published in a German anatomy and physiology journal in 1870 under the (translated) title: "On the Electrical Excitability of the Cerebrum."[4] Their results were important in that they had not only identified the motor area, albeit in an animal, but they had also confirmed, following Broca, that there was indeed localization of function in the cortex. In addition, the results suggested that electrical stimulation might prove a useful tool in demonstrating functions not only in animal brains but perhaps in brains of human subjects too.

One of those who took notice of the Fritsch and Hitzig publication was a young Scottish physician, David Ferrier (1843–1928; Figure 6.3). Keen to undertake some experiments of his own, Ferrier took advantage of the offer of good laboratory facilities in a rather unusual setting, the West Riding Lunatic Asylum in Yorkshire, England. Here, free from the distractions of a city university, Ferrier

SIR DAVID FERRIER, M.D.
1843-1928

Figure 6.3. Sir David Ferrier, Britain's first true neurophysiologist, after whom one of the Royal Society's most prestigious lectures is named. (Wellcome Collection CC BY)

embarked on a series of experiments in a variety of animals, though his subsequent studies on monkeys were performed in London at the Brown Institute. In most of the animals the muscle responses to electrical stimulation of the brain were examined and then the effects of removing part of the cortex were noted. Like other British physicians with an interest in neurology, Ferrier had been influenced by the doyen of that emerging specialty, John Hughlings Jackson (Box 6.1). It was Jackson who had proposed that there must be an orderly representation of movements somewhere in the brain since, in his patients with epilepsy,

Box 6.1 John Hughlings Jackson (1835–1911)

John Jackson was born in Yorkshire, England, the son of a brewer; he added "Hughlings" to his otherwise common name to give the latter more weight. In his late teens Jackson attended the York Medical and Surgical School before completing his medical training, aged 20, at St. Bartholomew's Hospital in London. After practicing in York for a year, Jackson returned to London and began the career that caused him to be regarded as the "father" of British neurology by his contemporaries. (The latter included Sir William Gowers [1845–1915] who, 10 years younger than Jackson, would now be judged by some as having made greater contributions to neurology, certainly to its clinical practice.) Important to Jackson were his appointments to various hospitals, especially the London Hospital and the National Hospital for Paralytics and Epileptics (later to become, largely through Jackson's influence, the National Hospital for Nervous Diseases, Queen Square). As there

Figure B6.1. John Hughlings Jackson. Venerated by his contemporaries and regarded as the "father" of British neurology. Photogravure after L. Calkin, 1895. (Wellcome Collection)

had been little previous advance in the diagnosis and understanding of neurological disorders, Jackson had to rely on his own observations and critical thinking, which he did to good effect. However, as his career progressed he continued to keep himself abreast of the publications of scientists working on the nervous system, especially those in France and Germany but also those of his friend, David Ferrier (see main text). Though appreciated by his peers, Jackson's speeches and writings are not easily understood, in part because of their prolixity. However, he continues to be remembered for his deduction (from the jerking movements preceding loss of consciousness in an epileptic patient) that there must be an orderly representation in the brain for the control of movement. Jackson also described the clinical features of temporal lobe epilepsy, particularly the entry into a "dreamy" state. Another powerful concept, novel at the time, was that the mammalian nervous system had evolved in "levels"—that is, a lower level (spinal cord and brainstem) had been supplanted by the development of a middle level (motor cortex) and that level, in turn, was to come under the influence of a higher level (the prefrontal cortex). Jackson was, in fact, a supporter of evolutionary theory and, at one point, had corresponded with Darwin over facial expression. Regarding consciousness, Jackson seems to have regarded this as a product of the brain operating as a machine.

he had repeatedly observed that a seizure would commence with involuntary jerking in one group of muscles and then spread to a neighboring one—for example, there could be a "march" from the muscles of the hand to those of the arm, shoulder, and neck. Where Jackson had been in error was in his suggestion of the striatum (the large nuclei at the base of the hemisphere), rather than the cortex, as the site of the movement map. Ferrier's experiments confirmed both Fritsch and Hitzig's claim of a cortical motor area and Hughlings Jackson's speculation of a sequential representation of movements. In the monkey the motor cortex was found adjacent to the central fissure; removal of this area produced hemiplegia on the opposite side of the body.[5] If more anterior areas were removed instead there were behavioral changes, though without any evidence of sensory or motor impairment—this last observation was consistent with the striking absence of symptoms in Phineas Gage (see previous discussion).

Though Ferrier's work was undoubtedly important and Ferrier himself would become one of the founders of the Physiological Society in Britain, it was later shown that there were two flaws in his work. One of these was in identifying, in primates, the strip of cortex behind the central sulcus as being part of the motor cortex; later work, by Sherrington and Leyton, would show that the primary motor area only involved the cortex in front of the sulcus. If that error could be explained by Ferrier's use of electrical stimuli that had caused the current to spread too widely in the cortex, there is no accounting for the second one—despite having apparently removed the occipital lobes in some of his monkeys, the animals were reported to have shown no evidence of blindness, unlike those that had undergone parietal ablations It was an error that would provoke strong criticism.

With the motor cortex having been mapped out in animals, it was only a matter of time before electrical stimulation was applied to the human brain. And, indeed, in 1874, only four years after Fritsch and Hitzig's publication, a Cincinnati surgeon, Roberts Bartholow (1831–1904), took advantage of the presence of exposed brain at the bottom of a large ulcer on the top of the head of a 30-year-old woman. By applying electrical stimuli through needles inserted into the cortex, he was able to identify an area that produced movements of the contralateral arm.[6] Later still, Charles Sherrington,[7] while occupying Richard Caton's former chair in physiology at Liverpool, undertook an extremely detailed study of the motor cortex in primates. With Albert Leyton (1869–1921) as co-investigator, no fewer than 22 chimpanzees, 3 gorillas, and 3 orangutans were examined under deep anesthesia, using weak faradic currents and a monopolar exploring electrode to enhance accuracy.[8] The gyrus immediately in front of the central sulcus (Rolandic fissure)[9] was confirmed as the primary motor cortex and, as David Ferrier had previously reported, the orderly representation of the movements was shown to be upside down, such that those for the leg were at the upper, medial border of the gyrus while those for the lips and tongue were at the lower, lateral end.

Leyton and Sherrington also noted that some parts of the body, such as the thumb, occupied much more of the cortex than others, such as the trunk and leg. Moreover, the regions overlapped, and the responses elicited by stimulating the same point tended to fluctuate. It was obvious to Leyton and Sherrington that the 400 different movements that they had been able to identify could be combined in different ways to serve different purposes. Unsurprisingly, the distorted and inverted map of movements in the primate brain was found to be essentially the same as that in human subjects when Wilder Penfield began to use stimulation during his neurosurgical procedures at the Montreal Neurological Institute in the 1930s.[10] However, unlike Leyton and Sherrington's study in primates, Penfield found that the central sulcus was not a firm boundary between motor and sensory responses. One of the legacies of Penfield's extensive work was the drawing of a motor homunculus overlying the cortex, a figurine that, in various forms, was to find its way into almost all subsequent neuroscience textbooks (Figure 6.4).[11] Interestingly, though Penfield is given credit

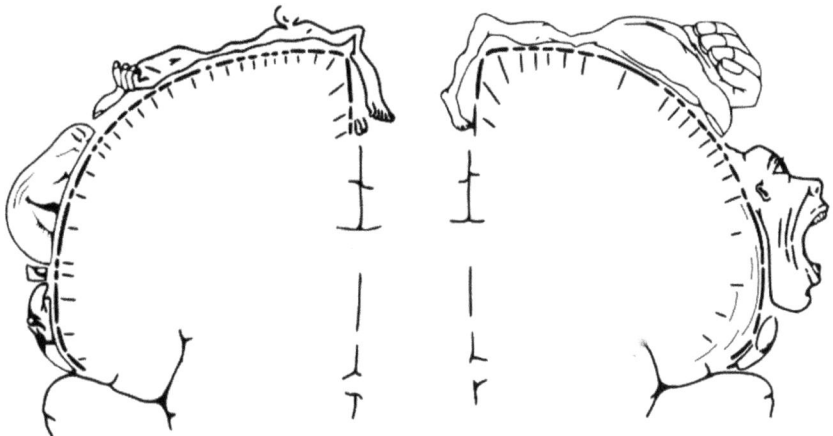

Figure 6.4. Representation of sensation (*left*) and of movements (*right*) in the postcentral and precentral gyri, respectively, as described by Penfield and Boldrey[10] and by Penfield and Rasmussen.[11] Both maps are simplifications because of the considerable overlaps of the body regions. (By mailto:ralf@ark.in-berlin.de [public domain], via Wikimedia Commons)

for pioneering systematic electrical stimulation of the human brain, his studies had been preceded by those of another neurosurgeon—Feodor Krause (1857–1937). Remembered as the "father" of neurosurgery in Germany, his homeland, Krause had already mapped out the motor cortex during the many operations for epilepsy he had performed prior to 1912 and he had published his findings. Penfield came to know this work during a sabbatical in Breslau in 1927 with Otfried Foerster, a disciple of Krause's and a practitioner of electrical mapping himself.

Among the later discoveries concerning the motor cortex was the finding that there were additional sites, close to the primary one, that also produced movements on electrical stimulation; these comprised a "supplementary" area on the dorsal border of the hemisphere, a "premotor" area just in front of the main area, and small areas on the medial side of the hemisphere. Also important, and clearly related to Leyton and Sherrington's observation of overlapping of movements in the cortex, was the demonstration by Charles Phillips (1916–1994) in Oxford that the same motoneuron in the spinal cord could be excited by stimulation over an appreciable region of cortex—evidently there were colonies of pyramidal neurons in the motor cortex that could call upon the same motoneuron for different purposes.[12]

Visual Cortex

So much for the localization of the motor regions of the cortex. What about the sensory areas? Here there were three possible approaches. One was to stimulate the human brain and ask the subject what he or she noticed; another was to stimulate a sensory pathway in an animal or human and to map out the resulting electrical potentials; while the third approach, again in both animals and humans, was to evaluate the sensory deficit after injury or disease involving the cortex. In the case of the visual cortex this last approach had been employed by Ferrier in his experiments on monkeys, though with misleading results—it was a lesion of the parietal lobe, rather than of the occipital lobe, that was reported to cause blindness. Ferrier's error was soon pointed out by his contemporary, Hermann Munk (1839–1912) at the Veterinary College in Berlin. Munk was able to demonstrate that removal of the occipital cortex on one side produced an inability to see objects on the opposite side, regardless of which eye was tested. This observation indicated, correctly, that both eyes projected to each side of the brain. Confident of his results, Munk was unsparing in his criticisms of Ferrier: "Mr Ferrier had not made one correct guess, all his statements have turned out to be wrong."[13]

Confirmation of Munk's conclusions would come from animal experiments by others and, perhaps more importantly, from studies of human patients who had suffered brain injury, particularly Japanese casualties in the 1905 war with Russia and British troops wounded in World War I. Both studies, the former by Inouye in Japan and the latter by Holmes and Lister in Britain, enabled the projection of the visual field onto the occipital cortex to be mapped out in detail. The key area was found to be the cortex surrounding the calcarine fissure on the medial side of each occipital lobe.[14]

At the same time that the neurologists were examining injured patients and the physiologists were starting to explore animal brains, the neuroanatomists were well advanced in their own studies of the cerebral cortex. Under the microscope it was evident that there were differences in the arrangements of neurons and fibers between one region and another. In the mesial (inner) side of the occipital lobe, for example, there was a very obvious white band of fibers running parallel to the cortical surface that was not apparent elsewhere—hence the term "striate" cortex. But how many anatomically defined cortical areas were there? Although different workers arrived at different numbers, the scheme that gained general acceptance was that of Korbinian Brodmann (1868–1918).

Brodmann (Figure 6.5) was born in Liggersdorf, a village in southern Germany close to the Swiss border. He attended several universities before qualifying in medicine in 1895. Having originally intended to become a general practitioner, Brodmann developed an interest in psychiatry and later in neuropathology. During a short period in Frankfurt he met Alois Alzheimer (of dementia fame), as well as several prominent neuroanatomists. By now resolved to make his career in research, Brodmann had little hesitation in joining Oskar and Cecile Vogt in the neurological institute that the Vogts had recently created in Berlin. It was in that city that Brodmann commenced his painstaking study of the cytoarchitecture of the mammalian cortex, comparing one region with another and doing so in a number of species. In 1909 his monograph appeared, and with it, the illustrations of the 47 areas he had been able to distinguish in the human brain (Figure 6.6).[15] It was a very considerable achievement and one that was destined to have a major influence on neuroscience. Yet, despite his success, and most likely because of his timid nature, Brodmann had difficulty in finding a senior position in a German university or research institute until shortly before his death (from septicemia in 1918). Regarding vision, Brodmann's area 17—anatomically, the "striate" cortex—was shown to coincide with the primary visual receiving area (V1) in the primate brain; it was the same part of the mesial occipital lobe identified as visual cortex by the neurologists.

What about the electrophysiologists—what was their contribution to the localization of vision in the brain? Richard Caton (Box 8.1), the discoverer of EEG activity in the brain, had made one pertinent observation in 1877 with his galvanometer readings from the brains of rabbits:

> A similar search was made to discover an area related to impressions on the retina. A point was found on the posterior and lateral part of the hemisphere in which, in three rabbits out of seven experimented on, variation of the current was seen to occur whenever a bright light was thrown upon the retina.[16]

However, it was not until the visual cortex had been identified by the anatomical and pathological studies that the electrophysiologists took over, first by recording

Figure 6.5. Korbinian Brodmann. (Wikimedia Commons)

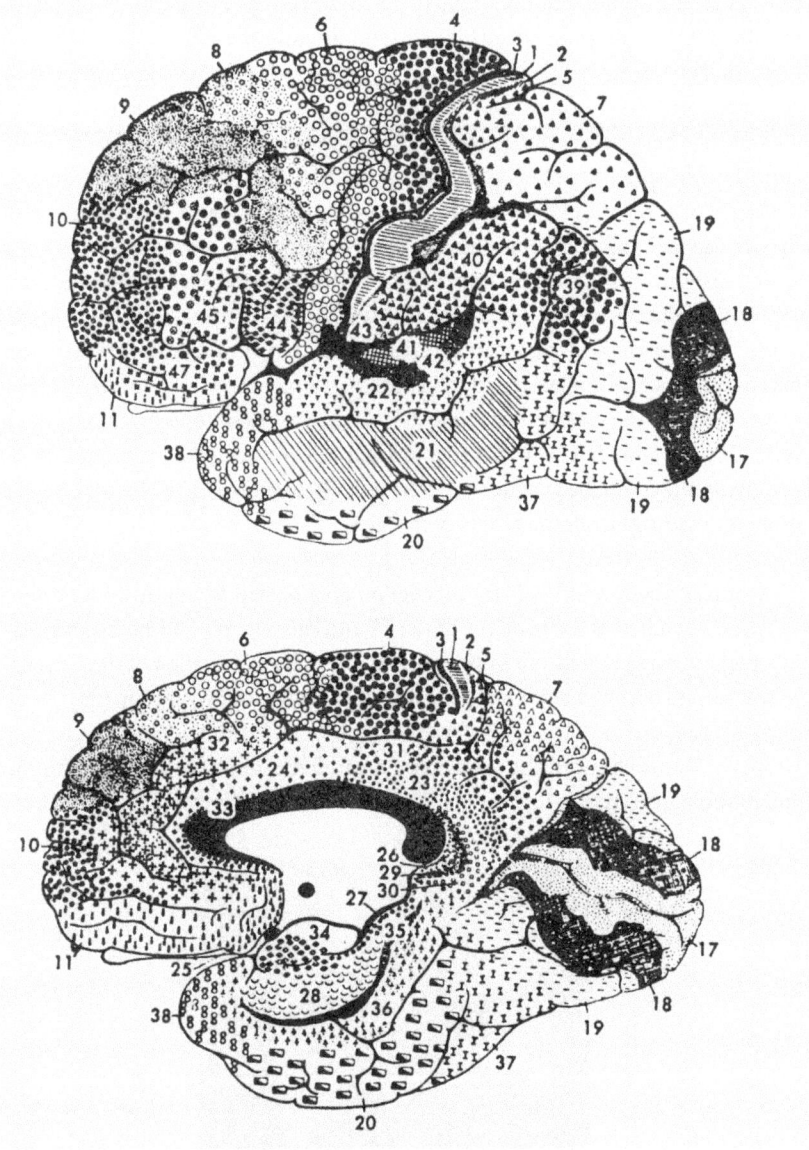

Figure 6.6. Brodmann's cytoarchitectural areas of the human cerebral cortex (upper and lower drawings show lateral and medial aspects, respectively).

the mass potentials evoked by flashing a light and then by ever more refined studies of single visual neurons, made with glass or metal microelectrodes. Prominent among the early investigators was George Bishop, the banished but intellectually adventurous Washington University physiologist who had been obliged to move his research laboratory to the pharmacology department. It was there that he and Howard Bartley employed electrical stimulation of the optic nerve to evoke slow-wave responses in the visual cortex of rabbits.[17] Another cortical pioneer was Ralph Gerard, the long-serving chair of the physiology department at the University of Chicago (Box 6.2). Rather than use electrodes on the surface of the cortex for his recordings, Gerard and his junior colleagues employed the same type of concentric needle electrode that Adrian had devised for studying electrical activity in contracting human muscles.[18] The advantage of such electrodes lay in

---

Box 6.2 Ralph Gerard (1900–1974)

Ralph Waldo Gerard was one of the giants of neurophysiology and, like Edgar Adrian in Cambridge and his own countryman George Bishop in St. Louis, there was hardly any part of the nervous system that he had not studied and made a significant contribution to. He was born in a small town in Illinois to a father who had graduated in engineering prior to emigrating to the United States from central Europe. Ralph was a precocious student, excelling in mathematics and science, and was able to graduate from high school and enter the University of Chicago at 15, obtain a PhD in physiology at 21, and qualify in medicine at 25. Declining to pursue a medical career, Gerard accepted a research fellowship that enabled him to study abroad with two Nobel laureates, A. V. Hill in London and Otto Meyerhof in Germany. In both cases Gerard investigated the biochemistry of peripheral nerve activity and, with Hill, successfully detected delayed heat production and, with Meyerhof, oxygen consumption. Returning to the United States, Gerard was first a lecturer and then, at a relatively young age, the head of physiology at the University of Chicago. In the latter position he was influential in training a number of noted neurophysiologists, among them Benjamin Libet. Gerard's prolific research included heat production in the active brain, brain mapping, spinal cord regeneration, DC potentials in the brain, and the contribution of summated slow potentials to EEG waves.

Gerard was the first to show that the transition from short-term to long-term memory involved protein synthesis. He was also the first to demonstrate the trophic effect of the nerve cell body on the maintenance of the axon and to show that this depended on the slow transport of protein along the axon. Ironically, Gerard is best remembered not for any of his remarkable discoveries but for developing, with Judith Graham, the glass capillary microelectrode—a tool that he used to measure the resting membrane potentials of muscle fibers but that became the neurophysiologists' primary resource in investigating the nervous system. In later years, Gerard was a spokesman for the role of science in society and in mental health research, particularly that of schizophrenia. Not surprisingly, considering his influence and achievements, Ralph Gerard was the recipient of numerous honors.

their superior ability to localize the source of impulse activity, especially when used in conjunction with a three-dimensional placement device (the Horsley-Clarke stereotaxic instrument) for investigating deep brain structures.

In relation to the visual pathway, the results from the various types of investigation, and from the later microelectrode studies in particular, led to an unexpected conclusion—a much larger area of cortex was devoted to vision than had been suspected. Thus, although the medial aspect of each occipital lobe received the projection from the retina, via the lateral geniculate body behind the thalamus, this primary cortical area, in turn, projected to other areas—not only in the occipital lobe but in the parietal and temporal lobes too. This wide distribution was the probable cause of Ferrier's mistake, in that the monkey with the parietal lobe lesion was not blind, as he thought; rather, its problem was that it could not make sense of what it saw. The large area of cortex devoted to vision tells us something else—humans, like most other mammals, are primarily visual creatures. What roles these various areas perform in the creation of the visual image is something that will be considered later.

Somatosensory Cortex

For somatic sensation, that is, the sensations referred to the external surface and interior of the body, some of the definitive work came from Wilder Penfield's operating room in Montreal. In patients undergoing surgery for epilepsy or tumor, he found that contralateral body sensations could be elicited by stimulation of the gyrus behind the central sulcus. Like the motor strip immediately in front, the representation was inverted, with the largest cortical areas devoted to sensations from the fingers and mouth (Figure 6.4). Essentially similar results were obtained by Clinton Woolsey and his colleagues at Johns Hopkins, where the experiments were conducted at about the same time as Penfield's but using the electrical potentials evoked in the monkey brain following stimulation of the skin.[19] Later, after moving to Wisconsin and having had contact with Penfield in Montreal, Woolsey carried out extensive mapping of human sensory cortex as well.[20] Both types of study, human and monkey, indicated that the figurine created for the motor cortex was equally suited to the somatosensory cortex.

Columnar Organization of Neurons in S1

Of the many scientists who came to study the activities of single neurons in the primary somatosensory receiving area, special mention must be made of Vernon Mountcastle (Box 6.3), who continued the work of the Woolsey group at Johns Hopkins. It was Mountcastle who, with Oxford's Tom Powell,

Box 6.3 Vernon Mountcastle (1918–2014)

Vernon Mountcastle combined all the best traits of a southern US gentleman with a dedication and ability that made him one of the most respected and admired neuroscientists of his generation—he was, in fact, elected the first president of the Society for Neuroscience by his peers. Born in Kentucky, Mountcastle grew up in Virginia, where his paternal grandfather had fought for the Confederacy in the Civil War; Mountcastle's father was a partner in a railroad construction firm and his mother a teacher. On completing college, this at the time of the Depression, Mountcastle was accepted by the Johns Hopkins Medical School in Baltimore—an association that was to last the better part of 80 years. After qualifying in 1943, Mountcastle began war-time service as a physician in the US Navy, taking part in both the Anzio landing in the Italian campaign and the D-Day invasion of Normandy.

With the ending of the war Mountcastle decided to specialize in neurosurgery, choosing to spend a year at Johns Hopkins studying neurophysiology first. Though the Department of Physiology at Hopkins, under the directorship of Philip Bard, had a small faculty, there was enthusiasm and excitement for the study of the brain—so much so that Mountcastle abandoned

Figure B6.3. Vernon Mountcastle (*left*) talking to Charles Phillips at the 1964 Cambridge meeting honoring Lord Adrian. (Courtesy of Professor Bill Harris, Department of Physiology, Development and Neuroscience, Cambridge)

his neurosurgical ambitions for the career of a neurophysiologist. Having started out with a project on the limbic system with Bard, Mountcastle then switched to the somatosensory system, and the latter became his subject for the rest of his working life. During this time he studied the responses of individual neurons sensitive to mechanical stimulation at all levels of the somatosensory pathway—peripheral nerve, thalamus, and postcentral cortex; the experiments were performed on monkeys, employed elaborate techniques, and involved very large numbers of observations. In the course of his work on the cortex he made the fundamental discovery that the cells functioned in columns (see main text). In the last part of his career, Mountcastle moved his microelectrode exploration of the cortex to that part of the parietal lobe behind the major somatosensory receiving area. It was there that he found neurons behaving as if they were directing attention and inducing motor responses to sensory stimuli—behaviors clearly related to conscious mechanisms and therefore of great interest.

Vernon Mountcastle lived to a great age (96) and during his life received as many honors and awards as a scientist might hope for, short of a Nobel prize.

determined the types of stimulus that were able to excite the cortical neurons; he also showed that skin receptors projected to cells that were flanked, in front and behind, by neurons responsive to muscle and joint receptors, so that the body was actually represented three times in the primary area.[21] Mountcastle also noticed something else—the neurons seemed to operate in columns perpendicular to the surface of the cortex. Thus all the neurons in one column would respond to stimulation of the same area of skin, while cells in an adjacent column would be excited from a neighboring patch. Mountcastle was generous in pointing out that cells functioning in columns had been a suggestion of Rafael Lorente de Nó (Box 6.4) many years previously, at a time when he had already carried out histological studies of the cortex.[22] Lorente had given his description in a chapter written for John Fulton in the latter's *Physiology of the Nervous System*, first published in 1938. The Spaniard, by then working in the United States, had taken great pains over his chapter, telling Fulton that "One of the reasons why writing it has been so laborious is that I have verified in my collection of brain sections the truthfulness of every statement in the text and of every line in the drawings." Much later, following Lorente and Mountcastle, David Hubel and Torsten Wiesel found that the columnar arrangement was also true of cells in the primary visual receiving area in the occipital lobe.[23] Here, then, was a very important clue as to how the cortex worked.

Box 6.4 Rafael Lorente de Nó (1902–1990)

Born in Zaragosa, Spain, Lorente de Nó would, in his lifetime, become recognized as the world's leading authority on the fine structure of the cerebral cortex, hippocampus, and brain stem, as well as a pioneering electrophysiologist. His intellectual abilities were manifested early, with the publication of a mathematical paper on thermodynamics at the age of 15. Attracted to medical science as a career, Lorente left Zaragosa to be taken on by Ramon y Cajal at a time when the latter was enjoying the fame of a Nobel Laureate and directing the research institute built for him in Madrid by the Spanish government. Though there were a number of talented neurohistologists already working under Cajal, Lorente was not only the youngest but was destined to become the most successful and widely known. During a three-year period in Robert Bárány's laboratory in Uppsala, Sweden, Lorente began a histological study of the vestibular nuclei in the brain. Returning to Cajal in Madrid, Lorente continued an exploration of the cerebral cortex. For a while, because of a dearth of research funds, he was obliged to earn his living by practicing medicine as an otolaryngologist. In 1931 he emigrated with his new wife to the United States, having accepted a research position at the Central Institute

Figure B6.4. Rafael Lorente de Nó—surely the most sophisticated neuroscientist of his generation and "rather too dashing" for Alan Hodgkin. (Courtesy of the Banco de imágenes de la Medicina Espanola)

for the Deaf in St. Louis. While in that city, he became interested in the developing field of neurophysiology; Joseph Erlanger and Herbert Gasser were at Washington University where, with George Bishop, they had pioneered the use of the cathode ray oscilloscope to study impulse conduction in peripheral nerve. Lorente built his own equipment and began to study the brain stem, interpreting his electrical recordings in terms of anatomical circuitry.

After several years in St. Louis, Lorente followed Gasser to the Rockefeller Institute in New York, where Gasser had been appointed director. While there, Lorente carried out an extremely thorough investigation of the electrical properties of peripheral nerve. It proved to be the one blemish in a very successful career, for his investigation of whole nerves was overtaken by Hodgkin and Huxley's elegant intracellular recordings from single giant axons of the squid. Worse, Lorente denied special roles for potassium and sodium ions in the resting and action potentials—though he correctly deduced the importance of oxidative metabolism in generating the resting potential (through the Na-K pump, as would be shown later). In his central nervous system research, on the other hand, Lorente had several notable achievements. In addition to his elaboration of the structure and function of the vestibular and auditory nuclei in the brain stem, there were descriptions of impulse activity proceeding in chains and rings of neurons in the central nervous system, the delineation of the main divisions of the mammalian hippocampus, the identification of cortical interneurons and interpretation of their roles, and—especially important—the recognition that neurons in the cerebral cortex were arranged in functional columns. More than anyone else at the time, it was Lorente who tried to link structure with function in the nervous system.

Higher Somatosensory Processing

Just as with the representations for movement and vision, it was found that there was more than one somatosensory area in the cortex, a secondary, smaller one having been found by Adrian in the upper lip of the lateral sulcus.[24] Little is known about the detailed function of this S2 area, originally discovered in the cat but also known to be present in primate brains, including those of humans. Part of the reason for this lack of information is the relative inaccessibility of the area, since the greater part is hidden from view. One important finding in the monkey, however, was that it seemed to be involved in manual discriminations—recognizing objects by feel.[25]

The cortex was discovered to contain another somatosensory area, however, though the territory was shared with vision. This was Brodmann's area 5, lying immediately behind the areas 3, 1, and 2 making up S1 in the postcentral cortex. There, in area 5, were neurons that responded to touch on the body surface and others to changes in joint position. The neurons differed from those in S1 in being more demanding—the stimuli were more complex. For example, some neurons would only fire to alterations in the angles of more than one joint. Other cells might respond to touch on the hand but only when the arm was flexed at the elbow or in some other critical position. A third class of cells would respond both to touch and to vision but in a linked manner. For example, if the touch area was on the head, then the effective visual stimulus would alter so as to always lie in front of the touch area, regardless of the position of the head at the time.[26]

The "Body Schema"

Both the electrophysiological recordings and the neurological syndromes occurring after damage to the brain suggested that the parietal lobe was responsible for creating a "body schema." This had been the term coined by the English neurologist, Henry Head, in the early 20th century.[27] Head had recognized that, without a subject being consciously aware of the positions of the various parts of his or her body at any given time, this information was continuously present in the brain in the form of a "schema." For example, if unaware of the body other than the hands, while sitting in front of a computer, one could immediately identify the back of the neck as the temporary landing ground for the annoying fly that had been buzzing round the room. In such an example, neural activity in S1 would have signified "touch" and Brodmann area 5 would have indicated "where." Another indication of the presence of a body schema would be the ability to reach down and scratch the back of the leg without having to look for either the arm or the leg.

Head envisaged the information from muscles and joints ("proprioception") as being responsible for his schema, but it then appeared, both from the neurophysiological recordings already described and from psychological experiments, that touch and vision were also involved. Vision, in fact, seemed the most important contributor to the body schema, as shown by the "rubber hand" experiment. In that test, the arm of a seated subject was hidden from view and a false arm substituted and positioned in front of him or her. If the visible false arm was then repeatedly touched, and the hidden arm touched at the same time, the subject came to accept the false arm as being the real one (Figure 6.7).[28] The ability of the visual system to compensate for the loss of sensory inputs from the skin, joints, and muscle was brought out by the well-documented case of a patient whose acute autoimmune inflammatory illness destroyed his sensory peripheral nerve fibers. Initially unable to sit up or move his limbs, it was only by dint of many hours of visually guided practice that the young man could regain his mobility.[29]

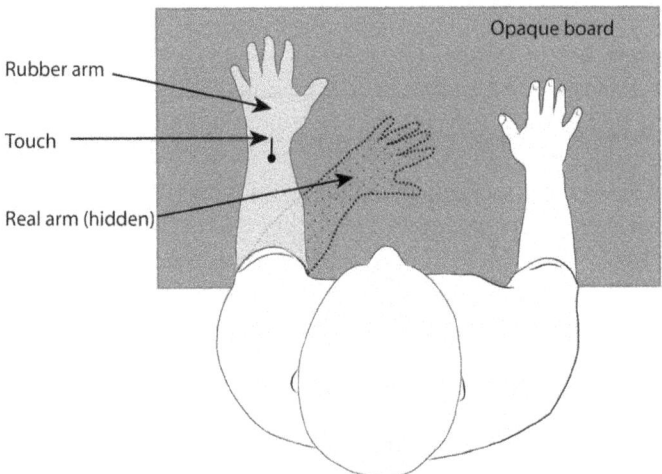

Figure 6.7. The rubber arm experiment. The subject comes to identify the visible rubber arm as his or her real arm when it and the hidden arm are touched simultaneously.

Clinical Distortions of the Body Schema

Given the unusual behavior of the neurons in this area of cortex (i.e., Brodmann area 5), it was perhaps not surprising that bizarre neurological syndromes could result from damage or overexcitation. Some of the most dramatic results were seen in a small minority of patients with migraine, either before or during the headache. Various parts of the body seemed enlarged or shrunken, arms and legs felt bent at impossible angles, and regions of the body might "disappear" (Figure 6.8). In patients with stroke or tumor, the aberrations were most marked if the damage involved the parietal lobe in the right hemisphere, since this lobe was entirely responsible for the left side of the body "schema," whereas both lobes contributed to the right side. Thus a patient who had a hemorrhage into the right parietal lobe might be totally unaware of the left side of the body. Alternatively he or she might deny that there was anything abnormal on that side, even if an arm or leg had been amputated.

That the spatial neglect extended beyond the body on the affected side and included vision was shown by the typical inability to draw the full face of a clock or even to recall both sides of a familiar scene. A striking example of the latter emerged when two Italian patients were asked to imagine themselves standing at one end of the *Piazza del Duomo*, the square in front of Milan's cathedral.[30] While they were able to name the buildings on the right side of the square but not those on the left, the situation was reversed when they pictured themselves standing at the other end of the square—clearly, all the buildings were in their memories, but the only ones that could be recalled depended on the direction they imagined themselves facing.

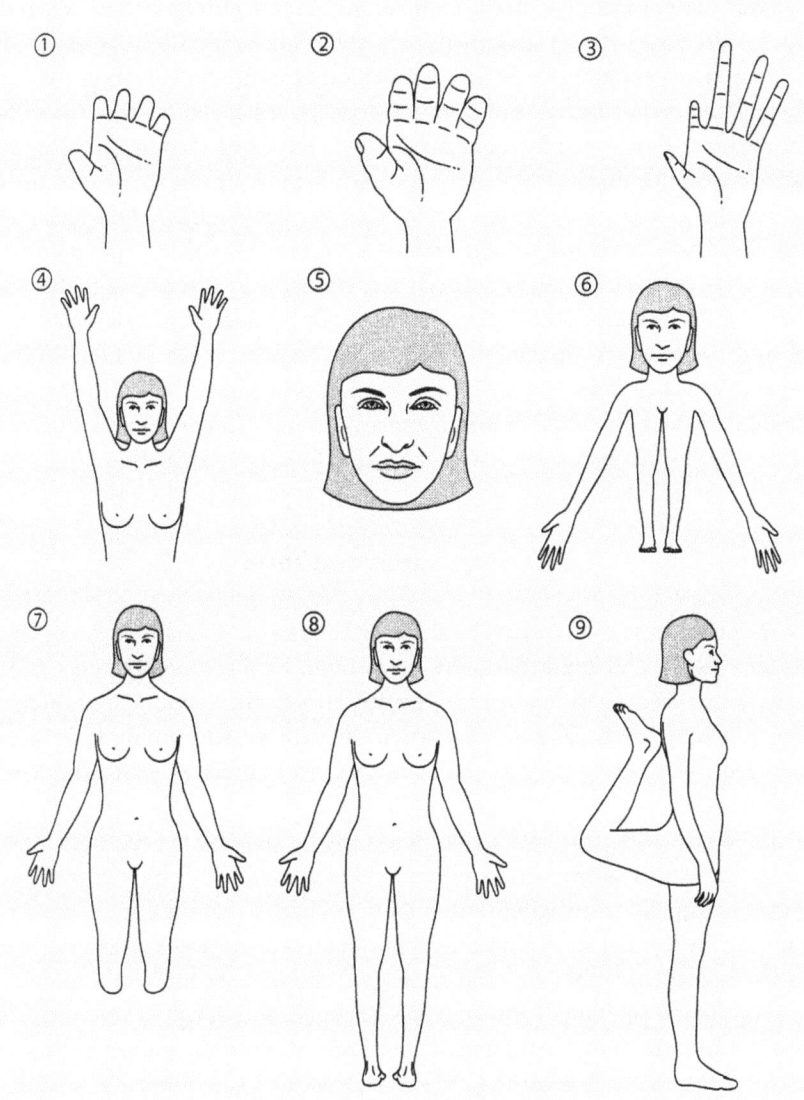

Figure 6.8. Distorted perceptions of the body in a patient with migraine, presumably due to overexcitation of the parietal lobe(s). (From: McComas AJ. The *artful chameleon: an exploration of migraine and medicine*. West Flamborough, ON: Alkat Neuroscience, 2006)

Area 7 and Motor Commands

It was the curious aberrations of spatial awareness, both of the body and outside the body, that prompted Vernon Mountcastle to explore the monkey parietal lobe with microelectrodes. In addition to the cells that required complex types of stimuli required to excite them, there were other neurons in area 7 that appeared to have a motor function, not only for directing gaze but also for arm movements. These cells fired before a movement took place and would do so before those in the primary motor area (M1).[31] Stimulation experiments in human subjects undergoing brain surgery reinforced the concept of parietal neurons directing movements. Thus, when the right inferior parietal region (Brodmann areas 39, 40) was weakly stimulated, there was a strong intention to move the hand, arm, or foot on the opposite side.[32] If the stimulus intensity was increased, the subjects felt that they had actually made the movements. In contrast, when the premotor region in the frontal lobe was stimulated the movements occurred, though this was denied by the subjects. Stimulation on the left side produced a similar dichotomy of intent and action, though confined to the muscles of the mouth. Various types of evidence, both anatomical and physiological, suggested that the parietal neurons in these areas, and in Brodmann areas 5 and 7, exerted their immediate effects via the premotor area in front of M1 (i.e., Brodmann area 6).

Hearing, Smell, and Taste

Of the other senses, the cortical location for hearing was found by a young physician in Russia. Vladimir Larionov, an assistant to Vladimir Bechterev in St. Petersburg, used a galvanometer to identify sites in the temporal gyri of dogs that produced electrical potentials when different-pitched tuning forks were sounded.[33] In human subjects the auditory cortex coincides with Brodmann areas 41, 42.

The final senses, smell (olfaction) and taste (gustation), are the ones about which the least is known—in part because they are more difficult to investigate. However, functional imaging studies have confirmed that the cortex most intimately involved in smell is in the hippocampus and adjacent entorhinal area in the medial temporal lobe—the same areas shown to be important in determining "place" in animals and concepts of people and things in humans.[34] In relation to taste, functional imaging and brain stimulation studies in human subjects, together with recordings of evoked responses in monkeys, showed that two cortical regions were involved, one in the insula and the other at the bottom of the frontal lobe (Brodmann area 43).

Comment: The Relationship of Motor and Sensory Areas to Consciousness

The identification of the motor and sensory areas account for rather more than half of the cortex. What is the contribution of these various regions to consciousness? This is a question that is considered in a later chapter, but it is convenient to make a comment now. In the case of the sensory regions, it is probable that the neural activity is the direct cause of the conscious experiences. Thus impulse firing in the visual cortical areas creates the image that is actually "seen" by the brain (though it may not be quite the same as the image recorded by the retina; see chapter 10). Similarly, impulse activity in the appropriate somatosensory areas may give rise to the feeling that something is touching the back of the neck, or, in another example, to the feeling of pressure in the finger tips while word processing. There is no need, nor any evidence, for some "higher center" in the brain to be involved.

The situation regarding the motor areas is very different, however, for there is good reason to believe that the primary, premotor, and supplementary areas cannot, by themselves, generate a conscious experience. A striking illustration of this inability comes from a well-known observation in clinical neurology in which a patient, while sleeping, develops a clot in the middle meningeal or internal carotid arteries. On waking the next morning, the patient attempts to switch off the alarm clock, only to find—to his or her surprise—that the arm will not move. Evidently the lack of blood to neurons in the motor areas in the middle of the contralateral hemisphere has prevented the execution of a movement ordered by some other part of the brain. Thus, while the cortical motor areas account for an appreciable part of the map of the cortex, these areas do not, by themselves, contribute to the conscious experience—something that Fritsch and Hitzig had foreseen in their 1870 paper and that had been confirmed much later by stimulating the human brain.

Memory

Short-Term Memory

One of the other functions that can be added to the map of the cortex is memory. To avoid confusion in a complex subject, one must first appreciate the different types of memory that have been recognized and that are referred to in neuroscience textbooks and papers. The basic distinction here is between short-term and long-term memories. *Short-term memory* is the ability to process new information and to hold it in the mind for a few minutes; this type of memory is also known as *working memory*. The ability to recall information that was processed longer than a few minutes beforehand is *long-term memory*. Although the distinction between these two types of memory may seem artificial, it is not; during short-term memory there is, in addition to any continuing impulse traffic, the potential for increased release of transmitter from the previously active nerve endings. At the same time structural changes at the synapses enable the information to be stored and possibly called upon later (i.e., long-term memory). A simple animal experiment by Ralph Gerard's group in Chicago demonstrated the transition from short term to long term. By ablating part of the cerebellum in rats they produced postural abnormalities in their hindlimbs; however, if the neural pathway between the cerebellum and the spinal cord was severed within 45 minutes of the ablation, the postural change did not occur.[35] Clearly, this critical time had been necessary for impulse activity to bring about the persisting effects at the synapses involved, as well as the creation of new synapses.

The elucidation of some of the molecular changes involved in the creation of memory was an achievement that gained a Nobel Prize for Eric Kandel at Columbia University in New York.[36] Kandel, like other leading neurophysiologists before him, chose an unusual but highly suitable preparation for most of his work—the sea slug, *Aplysia*. In some of his later studies, carried out in tissue culture, Kandel observed, under the microscope, stimulated *Aplysia* neurons sending out fine filamentous processes to form new synaptic connections. Kandel's earlier experiments, less rewarding ones, had been on the mammalian hippocampus; this part of the mesial temporal lobe was already known from clinical studies to be essential for creating explicit memory (see later discussion). Indeed, although persisting synaptic changes following activity could occur elsewhere in the nervous system, they were shown to be especially marked in the hippocampus by Timothy Bliss and Terje Lømo in Oslo. By repeatedly tetanizing the incoming perforant fibers, Bliss and Lømo showed that the responses of the hippocampal neurons remained greatly enlarged for several hours, an effect they termed "long-term potentiation." [37]

Long-Term Memory

There are two types of long-term memory, explicit and implicit. *Implicit memory*, also known as *procedural memory*, is the continued ability to perform actions that were learned in the past—riding a bicycle or playing the piano, for example. *Explicit memory*, also known as *declarative memory*, may be *episodic (autobiographical)* or *semantic*. *Episodic memories* are recollections of events, especially to the individual—a serious accident, for example, or the first dance with one's future spouse. *Semantic memories* refer to facts that can be recalled, such as the name of a politician or a capital city.

Once again, just as with the localization of motor and sensory areas, some of the clues for memory came from the neurosurgical operating theatre in Montreal where Wilder Penfield stimulated the exposed brains of his patients. It was when the stimulating electrode was in contact with the temporal lobe that some of his patients unexpectedly experienced events from the past. One such example was a 26-year-old woman who had suffered from seizures since the age of 5. Penfield found that stimulating a point in the anterior part of her superior temporal sulcus produced the following striking comments:

> "Yes, I heard voices down along the river somewhere—a man's voice and a woman's voice, calling." When asked what river, she said: "I do not know, it seems to be one I was visiting when I was a child." On repeating the stimulation, without warning, a different, but equally vivid memory was elicited: "Yes, I hear voices, it is late at night, around the carnival somewhere—some sort of travelling circus." After being asked what she saw, she said "I just saw lots of big wagons that they use to haul animals in."[38]

Neurological patients have provided other evidence for the importance of the temporal lobe, not just for the recall of past events but in the laying down of memory. The case of Patient HM (Henry Molaison) is especially well known. At the age of 27 this young man underwent resection of much of both medial temporal lobes—an operation never previously performed—by a leading neurosurgeon in Hartford, Connecticut.[39] The surgery was an attempt to relieve HM's long-standing epilepsy, and in this it was largely successful. Unfortunately, the young man was found to have become quite incapable of committing new events and facts to long-term memory or, indeed, to hold anything in his mind for more than a minute or so, though he remained able to perform tasks and even to learn fresh visuomotor skills. In effect, he had lost the capacity for *explicit* memory, though his *procedural* (*implicit*) memory was unimpaired. Increasing evidence indicates that the reason for a discrepancy of this kind is that *procedural* memory has more to do with the cerebellum than the cerebral cortex.

Sparing of *procedural* memory in the presence of severe *short-term memory* loss has been a feature of another notable patient, the Englishman CW (Clive Wearing).[40] At the age of 46, CW had been at the height of his career, having founded a choir that had been broadcasted and recorded, and being an international authority on late Renaissance music. He then contracted an acute inflammation of the brain caused by the herpes simplex virus and, typical of this disorder, had especially severe involvement of both temporal lobes. Afterwards, this highly accomplished and intelligent man was only able to remember facts, faces, and places for less than a minute, one of the results being that he greeted his wife as a stranger each day. To keep his sanity, CW kept a written account of every happening in the course of the day, appearing to "wake up" and become conscious every 20 seconds or so. Yet despite the almost complete loss of *short-term* (and *long-term*) memory, CW could still conduct and play music previously learned, indicating that his *procedural* memory was largely intact.

Update: The Human Brain Connectome Project

This account emphasizes that, from a combination of basic scientific and clinical observations, much was known about the activities of different brain areas long before the start of the present millennium. Nevertheless, in 2012 a major multicenter initiative in brain mapping began under the leadership of David Van Essen (Washington University, St. Louis)—the Human Brain Connectome Project. Previously Van Essen had been prominent in working out the cortical connections of the primary visual receiving areas, and the intention of the new project has been to extend this work by building a comprehensive map of neuronal connections in the entire human brain. To this end subjects, including identical twins, have been recruited and undergone studies with structural and functional magnetic resonance imaging (MRI), as well as density MRI (for showing nerve fiber tracts) and magnetoencephalography—the latter having excellent resolution for the timing of events though much less accuracy for their cortical locations. One of the major problems with such an ambitious study is that the topography of the cortical surface—the pattern of folding into gyri—varies markedly between subjects, even between identical twins. Nevertheless, early results from the project suggest that the number of functionally and connectively distinct areas is well in excess of the 47 determined for the human brain a century earlier by Brodmann.

Plasticity

The last issue to be considered is the durability of the brain maps—do the maps stay the same or, like the countries of the world, are there circumstances in which the borders change? The latter possibility was raised, among others, by Adrian and Matthews more than 80 years ago in their exploration of the "Berger rhythm"—the spontaneous 8–12 Hz EEG potentials that can be recorded over the back of the head when the eyes are closed. Noting the absence of the rhythm in patients who were blind, the authors made the following speculation about the visual cortex: "if vision is permanently cut off the area is not allowed to remain idle but becomes gradually more and more accessible to excitation from other parts."[41] It was a remarkably prescient remark and one that exemplified Adrian's uncanny insight into the workings of the nervous system.

That the question of plasticity should have been posed, however, suggests the answer—there are indeed situations in which a cortical area can acquire a new function. This mutability is not so surprising, however, given the way that the fetal cortex develops. Thus the occipital lobe is not genetically committed to processing visual information, nor even, in the frontal lobe, is the precentral gyrus obliged to elaborate instructions for movement. Instead, it is the cells in the thalamus and the associated geniculate bodies that will, once their axons have reached the cortex, determine the functions of the different regions. This commitment was suggested by various types of experiments. One, performed in tissue culture by Colin Blakemore (Box 6.5) and Zoltán Molnár in Oxford,

---

Box 6.5 Sir Colin Blakemore (1944–)

Colin Blakemore is not only an accomplished neuroscientist but, more than anyone else of his generation, the most prominent and influential spokesperson for neuroscience, and perhaps for science in general. Born in Stratford on Avon, Blakemore won a scholarship to Cambridge where he gained firstclass honors in medical sciences. Following PhD studies in vision at the University of California, Berkeley, Blakemore returned to Cambridge to

Figure B6.5. Colin Blakemore (*right*) with Jack Diamond at McMaster University, 1988. (Author's photograph)

become demonstrator and then lecturer in physiology. Pursuing his interest in vision, he was able to demonstrate, in kittens, a critical period for the development of synaptic connections within the visual cortex. Later studies were concerned with neural plasticity and included the observation that, in the developing brain, it was the outgrowth of fibers from the thalamus that largely determined the structure and function of the recipient cortical area. These last studies were carried out in Oxford University, where Blakemore had been appointed Waynflete Professor of Physiology at the young age of 35—it was the same position that Charles Sherrington had held more than 40 years before.

Because of his skill as a communicator, Blakemore was invited to give the BBC's prestigious Reith Lectures in 1976, choosing "Mechanics of the Mind" as their title. Thereafter he was much in demand as a lecturer and, in his trips around the world, would repeatedly fill the largest lecture halls when talking about neuroscience. While departmental head at Oxford, Blakemore initiated annual summer schools in neuroscience, intended for junior scientists and at no cost to those attending. It was during this period in his career that Blakemore incurred the enmity of the British animal rights movement. Almost alone among scientists, he had been prepared to argue the case for animal experimentation, but his courageous stand led to threats against his own life and those of his family.

Having proven himself an able administrator as well as a gifted neuroscientist, Blakemore was next appointed chief executive of the British Medical Research Council. Remarkably, but in keeping with his energy and principles, he was able to combine this position with research at Oxford. Knighted in 2014 (an overdue honor), Blakemore is now director of the Institute of Philosophy's Centre for the Study of the Senses at the School of Advanced Study in London. The recipient of numerous awards, prizes, and honorary memberships, Blakemore has many notable publications. Among these, his Harveian Oration, given in 2005 to the Royal College of Physicians, deserves to be read by every neuroscientist.

showed that fetal cortex that would normally become visual could be innervated by axons from the motor (ventrolateral) nucleus of the thalamus instead.[42] Indeed, any region of cortex could form functional connections with almost any thalamic nucleus. Another approach, this time in neonatal rats, has been to transplant cortex and to show, for example, that "visual" cortex can develop into motor cortex (Figure 6.9).[43] In a third type of experiment lesions were made in the visual pathways of newborn hamsters and ferrets, such that optic nerve fibers came to be connected, through a relay in the thalamus, to cortical regions that would normally comprise either the somatosensory or the auditory receiving areas. After further development of the operated animals it was possible to show

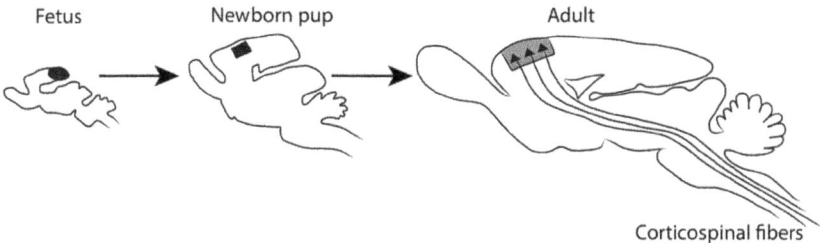

Figure 6.9. A piece of cortex is taken from the back of the fetal rat brain (future visual cortex) and transplanted anteriorly in a newborn rat pup. In their novel location the neurons develop axons running in the corticospinal tract—typical of the motor cortex.

that the "somatosensory" and "auditory" areas had now become visual ones instead, giving normal responses to visual stimuli.[44] The fetal cortex, then, is a *tabula rasa*—a blank slate waiting to be drawn upon, in this case by the axons from the thalamus.

It could still be argued, however, that, once the connections between the thalamus and cortex are formed, the commitment of the cortical area is final—once a "face" area, always a "face" area, for example. An early indication that this might not be so came from observations on patients with hand amputations, in whom evoked potential studies suggested that some of the cells in the primary somatosensory area representing the arm had become "leg" cells instead.[45] Such patients could also become aware of hand sensations while shaving—pointing to another invasion of the arm area, in this case by the axons and synapses for "face."[46] A third example of this *neuroplasticity* was the enlargement of the motor area for movements of the index finger in blind subjects proficient in reading Braille.[47] Additional experiments, using local anesthesia or ischemia in healthy subjects, have shown that some of the representational changes occur very quickly, presumably by the unmasking of previously ineffective synapses. Indeed, this rapidity might have been anticipated from the classic study of the primate motor cortex by Leyton and Sherrington, already mentioned;[8] these workers had noticed that careful stimulation of the same point on the cortical surface could give varying muscle responses.

In reviewing the many examples, two of the most convincing instances of the ability of a cortical area to change its function, in this case from one sense to another, have come from the pioneering studies of the late Paul Bach y Rita (1934–2006) in Wisconsin and the recent work of Amir Amedi in Jerusalem. Born in New York, Bach y Rita obtained his medical degree at the National University in Mexico City. After returning to the United States, he spent 10 years at the Smith-Kettlewell Institute of Vision Science in San Francisco before accepting a full professorial appointment at the University of Wisconsin in Madison. By that time he had already looked into the possibility of sensory substitution, initially using touch location on the skin of the back to provide information about surroundings to a blind person. To a large extent Bach y Rita's motivation came from witnessing the remarkably full recovery of his father who had been seriously disabled following a stroke. In a later development Bach y Rita used an array of stimulating electrodes on the tongue, in conjunction with an accelerometer on the head, to enable a patient to regain her sense of balance after her vestibular apparatus had been destroyed.[48]

Amedi's contribution to the field has been to show that congenitally blind subjects can be trained to sense objects, colors, and even facial expressions using auditory cues; in such patients functional brain imaging reveals neural activity spreading backwards from auditory cortex in the temporal lobe to "visual" cortex in the occipital lobe.[49] Neuroplasticity, then, is an important property of the cerebral cortex; fortunately it is also one that can be called upon, as in Bach y Rita's and Amedi's patients, to mitigate the effects of neurological disorder or injury.[50] There is, however, one cautionary note. Although a sensory modality can be made to occupy a foreign area of cortex, this does not necessarily mean that individual neurons have changed their function. There is the possibility that new "gnostic units" have been created from previously uncommitted neurons in the invaded cortical area.[51]

In summary, the task of mapping the cortex has been long, complex, and often difficult. Inevitably, as further information accrues, the map will be refined. For now, however, Figure 6.10 best sums up the present conception.

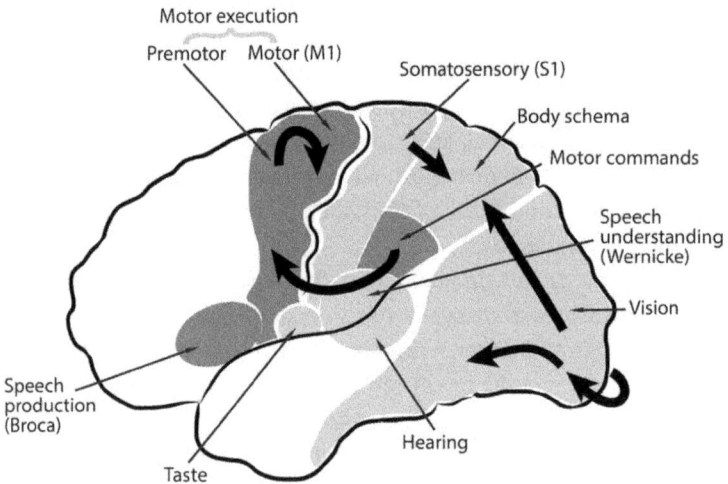

Figure 6.10. Summary of functions mainly associated with brain areas.

Next we move on to the excitation of the cortex. In Sherrington's analogy, what are the factors that cause the myriad points of light to twinkle and the shuttle to speed across the loom?

Notes and References

1. Crick FC, Koch C. What is the function of the claustrum? *Philosophical Transactions of the Royal Society B*, 2005; 360 (1458): 1271–1279. (The claustrum is a thin sheet of neurons deep to the cortical layers.)
2. Dennett D. *Consciousness explained*. Boston, MA: Little Brown, 1991.
3. The history of early research into the function of the cerebral hemispheres is well documented in the article by Fritsch and Hitzig (see note 4).
4. Fritsch F, Hitzig E. Ueber die elektrische Erregbarkeit des Grosshirns. *Archiv für anatomie, physiologie und wissenschaftliche medizin*. 1870; 37: 300–332. Translated by Wilkins RH. Neurosurgical classic—XII. *Journal of Neurosurgery*. 1963; 20: 904–916.
5. Ferrier D. The Croonian Lecture. Experiments on the brain of monkeys (second series). *Proceedings of the Royal Society of London*. 1875; 165: 433–488.
6. Bartholomow, R. Investigations into the functions of the human brain. *American Journal of the Medical Sciences*. 1874; 66: 305–313.
7. See p. xv for information concerning Sherrington.
8. Leyton ASF, Sherrington CS. Observations on the excitable cortex of the chimpanzee, orang-utan and gorilla. *Quarterly Journal of Experimental Physiology*. 1917; 11: 135–222.
9. See Figure 3.2.
10. For information on Penfield, see Box 3.2. The first full paper was: Penfield W, Boldrey E. Somatic motor and sensory representation in the cerebral cortex of man as studied by electrical stimulation. *Brain*. 1937; 60: 389–443.
11. Penfield W, Rasmussen T. *The cerebral cortex of man*. New York, NY: Macmillan, 1950.
12. Phillips CG. The Ferrier Lecture, 1968. Motor apparatus of the baboon's hand. *Proceedings of the Royal Society of London B*. 1969; 173: 141–174.
13. Cited in Note 14.
14. Glickstein M. The discovery of the visual cortex. *Scientific American*. 1988; 258: 118–127.
15. Brodmann K. *Vergleichende lokalisationlehre der grosshirnrinde*. Leipzig, Germany: Barth-Verlag, 1909. This important monograph, of which few copies exist, has been translated: Garey L. *Brodmann's localization in the cerebral cortex: the principles of comparative localization in the cerebral cortex based on cytoarchitectonics*, 3rd ed. Boston, MA: Springer, 2006. For the background to this work, see: Loukas M, Pennell C, Groat C, Tubbs RS, Cohen-Gadol AA. Korbinian Brodmann (1868–1918) and his contributions to mapping the cerebral cortex. *Neurosurgery*. 2011; 68: 6–11.
16. Caton R. Electrical currents in the brain. *British Medical Journal*. 1875; 2: 278.
17. Bartley SH, Bishop GH. The cortical response to stimulation of the optic nerve in the rabbit. *American Journal of Physiology*. 1932; 103: 159–172.

18. Gerard RW, Marshall WH, Saul LJ. Electrical activity of the cat's brain. *Archives of Neurology and Psychiatry*. 1936; 36: 675–735.
19. Woolsey had come to use evoked potentials rather than ablations for mapping because of the success of Gerard's group in Chicago and the move to Hopkins of one of Gerard's team, Wade Marshall, with the necessary equipment.
20. Woolsey CN, Erickson TC, Gilson WE. Localization in somatic sensory and motor areas of human cortex as determined by direct recording and electrical stimulation. *Journal of Neurosurgery*. 1979; 51: 476–506.
21. Powell TP, Mountcastle VB. Some aspects of the functional organization of the cortex of the postcentral gyrus of the monkey: a correlation of findings obtained in a single unit analysis with cytoarchitecture. *Bulletin of the Johns Hopkins Hospital*. 1959; 105: 133–162.
22. Lorente de Nó R. The cerebral cortex: architecture, intracortical connections and motor projections. In: Fulton JF. *Physiology of the nervous system*. London, England: Oxford University Press, 1938:291–325. See Figure 7.4 in the next chapter for a summary of Lorente's observations.
23. See Chapter 10.
24. Adrian ED. Double representation of the foot in the sensory cortex of the cat. *Journal of Physiology*. 1940; 98: 16P–18P.
25. Garcha HS, Ettlinger G. Tactile discrimination learning in the monkey: the effects of unilateral or bilateral removals of the second somatosensory cortex (SII). *Cortex*. 1980; 16: 397–412.
26. For review of the field, see Graziano MSA, Botvinick MM. How the brain represents the body: insights from neurophysiology and psychology. https://www.princeton.edu/~graziano/Papers/Attn_Perf19.pdf
27. For more on Henry Head, see Box 3.1. For the concept of the body schema, see Head H. *Studies in neurology*, Vol. 1 and 2. London, England: Henry Frowde, 1920.
28. Botvinick M, Cohen JD. Rubber hand "feels" what eye sees. *Nature*. 1998; 391: 756.
29. Cole J. *Pride and a daily marathon*. Cambridge, MA: MIT Press, 1995.
30. Bisiach E, Luzzati C. Unilateral neglect of representational space. *Cortex*. 1978; 14: 129–133.
31. Mountcastle V. Brain mechanisms for directed attention. *Journal of the Royal Society of Medicine*. 1978; 71: 14–28.
32. Desmurget M, Reilly KT, Richard N, Szathmari A, Mottolese C, Sirigu A. Movement intention after parietal cortex stimulation in humans. *Science*. 2009; 324: 811–813.
33. Larionow W. Ueber die musikalischen centren des gehirns. *Archiv für die Gesamte Physiologie des Menschen und der Tiere*. 1899; 76: 608–625.
34. See Chapter 9.
35. Chamberlain TJ, Halick P, Gerard RW. Fixation of experience in the rat spinal cord. *Journal of Neurophysiology*. 1962; 26: 662–673.
36. Kandel's research into the cellular mechanisms involved in creating memory, much of which was conducted on the sea slug, *Aplysia*, is summarized in his Nobel Lecture; the latter is easily found on the Internet through the Nobel website (Nobelprize.org). Kandel's own life story can be found in his entertaining autobiography, *In search of*

memory: the emergence of a new science. New York, NY: W. W. Norton, 2006. A more up-to-date account of the molecular mechanisms involved in memory formation is given in: Kandel ER. The molecular biology of memory: cAMP, PKA, CRE, CREB-1, CREB-2, and CPEB. *Molecular Brain.* 2012; 5: 14. https://doi.org/10.1186/1756-6606-5-14.

37. Bliss T, Lømo T. Long-lasting potentiation of synaptic transmission in the dentate area of the anaesthetized rabbit following stimulation of the perforant path. *Journal of Physiology.* 1973; 232: 331–356. For the background to this seminal work, see Lømo T. The discovery of long-term potentiation. *Philosophical Transactions of the Royal Society of London B.* 2003; 358: 617–620.
38. Penfield W, Perot P. The brain's record of auditory and visual experience: a final summary and discussion. *Brain.* 1963; 86: 595–651.
39. HM has been the subject of numerous papers, mostly by Brenda Milner (McGill University) and Susan Corkin (MIT). However, the best read for HM's entire story is the recent book by his neurosurgeon's grandson: Dittrich L. *Patient H.M. A Story of memory, madness and family secrets.* London, England: Chatto & Windus, 2016.
40. YouTube videos of Clive Wearing, both informative and touching, are available on the Internet. That Clive's intelligence and personality should have survived his terrible handicap is due, in part, to the daily attention of his wife, Deborah, seen with him in the videos.
41. Adrian ED, Matthews BHC. The Berger rhythm: potential changes from the occipital lobes in man. *Brain.* 1934; 57: 355–385.
42. Molnár Z, Blakemore C. How do thalamic axons find their way to the cortex? *Trends in Neurosciences.* 1995; 18: 389–396.
43. O'Leary DM, Stanfield BB. Selective elimination of axons extended by developing cortical neurons is dependent on regional locale: experiments utilizing fetal cortical transplants. *Journal of Neuroscience.* 1989; 9(7): 2230–2246.
44. Métin C, Frost DO. Visual responses of neurons in somatosensory cortex of hamsters with experimentally induced retinal projections to somatosensory thalamus. *Proceedings of the National Academy of Sciences of the United States of America.* 1989; 86: 357–361.
45. Sica REP, Panizza M, Reich E, Correale J. Modifications of the N1-P1 component of the somatosensory evoked potential in humans after partial limb amputation as a manifestation of CNS remodeling. *Electromyography and Clinical Neurophysiology.* 1988; 28: 227–231.
46. Ramachandran VS, Stewart M, Rogers-Ramachandran DC. Perceptual correlates of massive cortical reorganization. *NeuroReport.* 1992; 3: 583–586.
47. Pascual-Leone A, Cammarota A, Wassermann EM, Brasil-Neto JP, Cohen LG, Hallett M. Modulation of motor cortical outputs to the reading hand of Braille readers. *Annals of Neurology.* 1993; 34: 33–37.
48. Danilov UP, Tyler ME, Skinner KL, Hogle RA, Bach-y-Rita P. Efficacy of electrotactile verstibular stimulation substitution in patients with peripheral and central vestibular loss. *Journal of Vestibular Research.* 2007; 17: 119–130.

49. Striem-Amit E, Cohen L, Dehaene S, Amedi A. Reading with sounds: sensory substitution selectively activates the visual word form area in the blind. *Neuron*. 2012; 76: 640–652.
50. Hallett M. Review. Plasticity of the human motor cortex and recovery from stroke. *Brain Research Reviews*. 2001; 36: 169–174.
51. See Chapter 12 for more information on gnostic units.

7
Awaking the Cortex

Bremer's Novel Brain Preparations

At times when serious crimes, political or religious differences, or even the failure to conceive a male heir to a throne were punishable by beheading, there must have been speculation among onlookers as to the fleeting thoughts and sensations of the detached brain of the victim. Though the answer to this macabre musing must forever remain conjecture,[1] it has been possible, from animal experiments, to learn something about function in a brain that has been divided while still retaining its blood supply. Though many, scientists included, might hesitate to condone experiments of this nature, at the time that they were undertaken little thought was given to the possibility of consciousness in animals.[2] Moreover, the person who began this line of experimentation was, in every other way, exemplary.

Frédéric Bremer (Box 7.1) was a Belgian who completed his training in neurology following World War I. Following postdoctoral studies in the United States and Britain, he began to combine his clinical responsibilities with original experiments in several areas of the nervous system, including those concerned with muscle tone. For some of the latter experiments it was usual to separate the part of the nervous system under study—the spinal cord—from the influence of the "higher centers." This isolation usually involved sectioning the brain stem and then removing the two cerebral hemispheres. Bremer, ever curious, wondered what sort of function might be found in the cerebral cortex if the hemispheres were left in place. Crucial to this issue was the normal function of the brain stem. At that time, most of the knowledge about this "stalk" concerned its anatomy rather than its physiology. From dissections it had long been known where the various cranial nerves exited (to the nose, eyes, face, mouth etc.); also known were the main fiber bundles running to and from the spinal cord, as well as the fact that the neurons controlling breathing were in the central substance (reticular formation) of the brain stem. But that was about all.

In his first experiments, carried out in the University of Brussels, Bremer sectioned the brain stem of a cat at the level of the midbrain, doing so under anesthesia and then placing recording electrodes on the surface of the cortex (Figure 7.1). Though the anesthesia was allowed to wear off, the constricted pupils and squint of the cat's eyes suggested that the animal, or rather its brain,

> **Box 7.1 Frédéric Bremer (1892–1982)**
>
> A brilliant yet modest man and a true neurological pioneer, Frédéric Bremer was born in Arlon, a small town in the forested Ardennes in the southern part of Belgium; both his parents were teachers in the local public school. Bremer was an excellent student and went on to the University of Brussels to study medicine; however, the outbreak of World War I was followed by the German invasion of Belgium, with disastrous consequences for that country. Bremer interrupted his studies to serve as a physician with a cavalry regiment and then as a medical auxiliary at a military hospital on the coast. After the war Bremer finished his medical studies and, having decided to specialize in neurology, went to Paris for training with Professor Pierre Marie at the famous La Salpêtrière hospital, the same hospital in which Charcot had taught and practiced. Bremer then went overseas to continue his postdoctoral studies, first in Boston with Harvey Cushing, the neurosurgical pioneer, and then in Oxford with Charles Sherrington, the eminent neurophysiologist.
>
> Returning to Belgium in 1924, Bremer joined the University of Brussels where he would become, in 1934, Professor of General Pathology. However, in addition to his departmental duties and his teaching, Bremer was a very able clinician and continued to see neurological patients. This was not all, however, because Bremer's strong scientific curiosity led to a long series of experiments into the neural mechanisms producing sleep. In the course of these studies Bremer devised two novel animal preparations, the cerveau isolé and the encéphale isolé (see main text). After World War II Bremer became active in research again, experimenting by himself or with visiting scientists and attempting to make a coherent synthesis of the various neural influences affecting sleep. The outcome of his labors was the fine review that appeared five years before his death.
>
> *Source.* Kerkhofs M, Lavie P. Frédéric Bremer 1892–1982: a pioneer in sleep research. *Sleep Medicine Reviews.* 2000; 4: 505–514. For photograph, see Figure 3.3.

was still asleep. The cortical recordings were also suggestive of sleep, or at least drowsiness, since they showed typical "sleep spindles"—that is, alternating potentials, with a frequency of 8 to 12 Hz, that waxed and waned. Bremer's interpretation of his findings was that, having been deprived of all sensory input from the body below the neck, as well as from the face, scalp, and mouth, the cortex had indeed gone to sleep. This, then, was the *cerveau isolé* (isolated cortex) preparation.

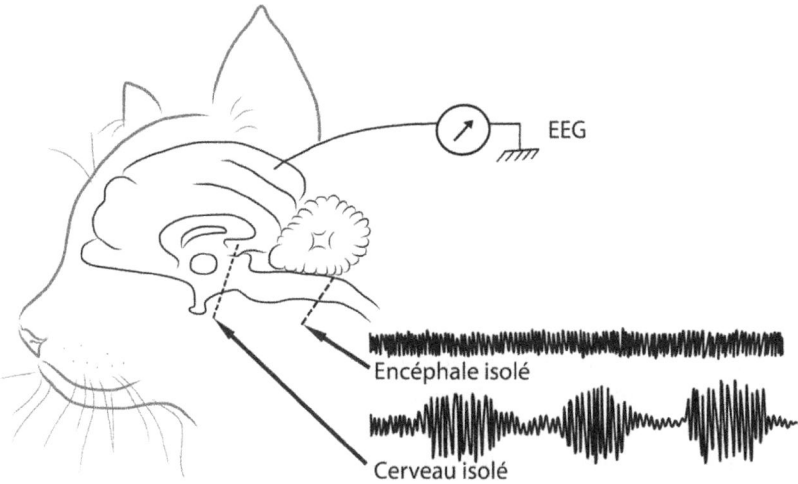

Figure 7.1. Bremer's two preparations, the sections through the brain stem being shown by interrupted lines. The EEG activities characteristic of the two preparations are shown schematically on the right (fast activity in the encéphale isolé and 8–12 Hz spindles in the cerveau isolé).

The publication describing these novel observations appeared in 1935[3] and was followed two years later by another paper, again with Bremer as its sole author and with equally fascinating results.[4] This time Bremer reported what he had observed if the brain was sectioned lower down, at its junction with the spinal cord. The cat's eyes now gave the appearance of wakefulness, and this interpretation was supported by the replacement of sleep spindles in the EEG with higher frequency, low amplitude activity (Figure 7.1). Bremer postulated that it was ongoing sensory information coming from the tissues of the head, via the trigeminal nerve and its brain stem nuclei, that was keeping the cortex alert; that information had been unavailable to the cortex in the cerveau isolé preparation. It was a plausible explanation of Bremer's novel findings and one that was accepted by the neuroscientific community.

The first hint that it was inadequate came from the other side of the Atlantic.

Nonspecific Thalamic Nuclei

With the outbreak of World War II in 1939, research unrelated to the war efforts of the participating countries dwindled, and this included neurophysiology in Britain and Europe. With their country entering the war later, American scientists were able to keep active for rather longer, and this was true in the neurophysiological laboratories at Harvard.

Originally founded in 1636 as a theological college, Harvard had become a university with widening interests, interests that would encompass science in the 1800s. As far as medicine was concerned, the Massachusetts Medical College had moved from the Harvard campus to the city of Boston in 1870, and in 1908 the Division of Medical Sciences was established, with its departments of anatomy and physiology. Despite its small size, the physiology department in the 1920s and 1930s was highly productive. One of its members was Hallowell Davis, destined to become the world's leading authority on the physiology of hearing. Davis had also made the first EEG recordings in the United States with a machine put together in his laboratory. Later, his engineer Albert Grass would start his own company, designing and building stimulators, amplifiers, EEG machines, and other equipment that would be used all over the world.

The head of the department of Harvard was a superb neurophysiologist too. Born into a wealthy family, Alexander Forbes (Figure 7.2) was able to combine an interest in neuroscience with the life of an adventurer—including sailing his yacht across the Atlantic and piloting his own plane. As a neuroscientist, he

Figure 7.2. Alexander Forbes (*right*) visiting Charles Sherrington (*center*) with Edgar Adrian. The photograph would have been taken in the backyard of Sherrington's house in Ipswich, Suffolk, in the late 1930s after Sherrington's retirement from Oxford.

visited Sherrington in Liverpool before World War I and was a friend and admirer of Edgar Adrian in Cambridge before and after that war. Forbes was the first to employ valve amplification to improve recordings from peripheral nerves[5] and he was, like Davis, Jasper, and Grey Walter, a pioneer in the development of the EEG. Aware that the United States might join in the war, Forbes offered his services to the US Navy and undertook aerial mapmaking of the Labrador coast, searching for sites that might prove suitable for advanced airfields. Had Forbes not done this, he would, as Harvard's senior neuroscientist, almost certainly have participated in some novel experiments that would shed considerable light on the functioning of the brain stem. As it was, he made the fundamental observations and it is likely that he had been involved in the planning of the later experiments.

The observation that Forbes had made with his assistant Morison[6] was that, when a peripheral nerve was stimulated in a deeply anesthetized animal, it was possible to record in the cortex not only a "primary" response, largest in the contralateral sensory receiving area, but also a later, widely dispersed, "secondary" discharge.[7] Forbes speculated about the origin of this unexpected late wave and thought that, while intercortical pathways might also be involved, "the afferent volley may act upon the thalamus or other subcortical centers, and that these centers may then distribute the discharge throughout the cortex."

With Forbes on leave from Harvard, however, it fell upon Robert Morison, another faculty member, to take the lead in the investigation of the possible subcortical pathway. Using a stereotaxic device of his own design, and with Edward Dempsey and other colleagues from the department, Morison was able to stimulate different areas of the thalamus and brain stem selectively and to see if the secondary discharge could be replicated. Morison also investigated the possible relationship between the secondary discharge and the spontaneous EEG activity in the cortex. The two main conclusions from this extensive body of work were that (a) in addition to the well-known somatosensory pathway through the medial lemniscus and ventral posterior thalamic nuclei, there was another route that passed through the "nonspecific" thalamic nuclei and was widely distributed in the cortex and (b) the neurons in this nonspecific projection were also responsible for generating spontaneous EEG spindles. The novel findings led to 10 original papers being published in a three-year period.[8,9,10]

The discovery of a nonspecific thalamocortical system of neurons was a considerable advance in brain physiology and was immediately recognized as such. With one major success behind him, Morison may well have looked forward to others. Instead he published his last neurophysiology paper in 1945 and, leaving Harvard for the Rockefeller Foundation, became, like so many other prominent scientists, an administrator.[11]

With Morison gone and the brain research at Harvard at a standstill, it was Herbert Jasper (Box 3.4) who took up the nature of the relationship between the nonspecific thalamic nuclei and the cortex. World War II had been a busy time for him; while still responsible for the clinical neurophysiology service at the Montreal Neurological Institute, Jasper had undertaken government work related to the war effort. As if that were not enough, he had also enrolled in the MD program at McGill University, relying on a young David Hubel for his student lecture notes. Now it was time for Jasper to make use of the animal laboratories that Penfield had created for him in the Neurological Institute and to have other neuroscientists, as well as graduate students and postdoctoral fellows, work with him. His first achievement, undertaken in collaboration with visiting colleagues, was to repeat the experiments of Morison and Dempsey and to confirm their findings.

Impressed by the powerful actions of the nonspecific thalamic nuclei, and mindful of Penfield's concept of a centrencephalic integrating system subserving consciousness, Jasper ended a presentation with "a central integrative mechanism . . . is necessary to explain consciously directed thought and behavior. It seems that the thalamic reticular system with its diffuse cortical projections . . . is a good candidate for this office."[12]

With Choh-Luh Li and Cullen, Jasper proceeded to explore the responses to thalamic stimulation in more detail; for this he used microelectrodes for recording in addition to the silver or wick electrodes on the surface of the brain. As the microelectrode penetrated the different layers of the cortex, he observed how the form of the electrical responses changed. When a specific thalamic nucleus was stimulated there was excitation in the deeper cortical layers first (Figure 7.3), but when the nonspecific nuclei were stimulated instead, there was excitation throughout all layers of the cortex.[13,14,15] Jasper related these findings to the drawings of cortical neurons and nerve fibers that had been made by the Spanish histologist and electrophysiologist Rafael Lorente de Nó some years before (Figure 7.4).[16]

Jasper drew one more conclusion from this later work, and, for understanding the possible mechanisms underlying consciousness, it was a very important one—he changed his mind about identifying the thalamic reticular system with Penfield's centrencephalic system: "We do not believe that one will find in the thalamus the system of neurons fulfilling the definition proposed by Penfield for a centrencephalic system." The reason for the change of heart was that, whereas Penfield envisaged his centrencephalic system as projecting to both cerebral hemispheres, Jasper showed that the thalamic nuclei only made connections with the cortex on the same side. Any bilateral effects would have to come through the corpus callosum, the massive bridge of nerve fibers connecting

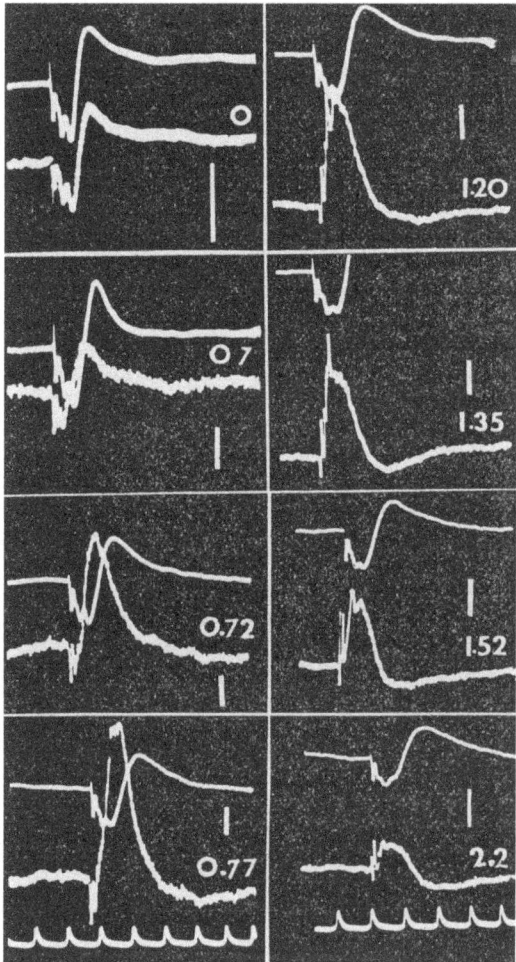

Figure 7.3. Potentials evoked in cat's somatosensory cortex following stimulation of the corresponding specific thalamic nucleus. Each of the eight pictures show recordings from the surface (*upper trace*) and interior of the cortex (*lower trace*; depth in mm indicated on the right). While the surface potential remains much the same, the interior potential reverses polarity and is maximal at a depth of 0.77 mm, indicating the location of a response in the cortex (depolarization of pyramidal cell apical dendrites in layer IV). The intervals on the time traces at the bottom of the figure correspond to 5 ms. The responses in this figure are examples of the "slow" waves discussed in the next chapter. (Reproduced with permission from: Li C-L, Cullen C, Jasper HH. Laminar microelectrode studies of specific somatosensory potential. *Journal of Neurophysiology.* 1956; 19: 111–130).

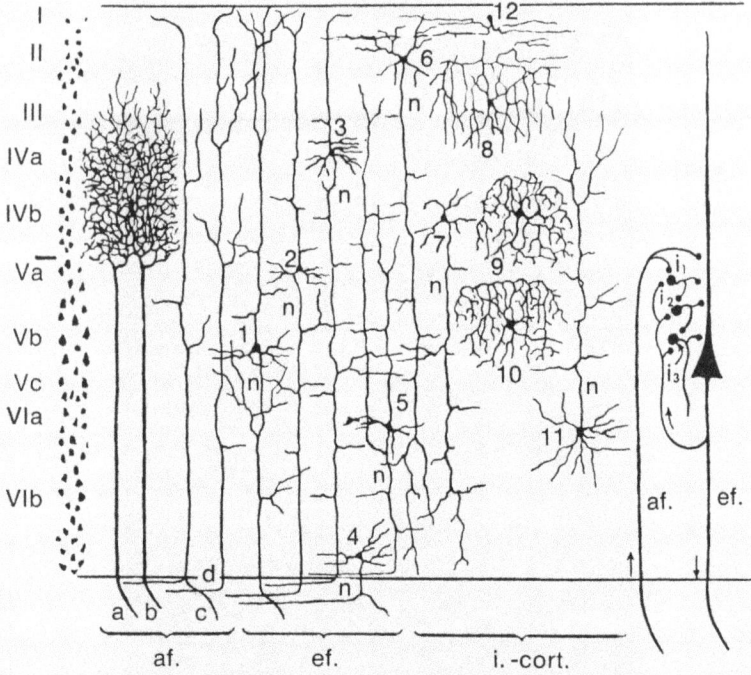

Figure 7.4. This diagram of the microscopic structure of the cortex is a composite made by Alf Brodal from several of Lorente de Nó's figures. On the far left are Roman numerals showing the six cortical layers and next to them are cell bodies of the pyramidal neurons forming two to three columns, with the largest cells in layer V. Moving to the right, typical pyramidal cells are numbered *1–3* and show the long apical dendrites reaching to the cortical surface; cells *8–10* are interneurons. Incoming axons (*a,b*) from specific thalamic nuclei terminate in dense arborizations in layer IV, while the endings of a nonspecific thalamic axon (*c*) are distributed more widely. (From: Brodal A. *Neurological anatomy in relation to clinical medicine*, 3rd ed. Oxford, England: Oxford University Press, 1981)

the two hemispheres, or from branching of ascending nerve fibers in the lower brain stem.

Discovery of the Reticular Activating System

At the same time that Jasper was conducting his experiments in Montreal, an important discovery was being made on the other side of the US–Canada border, in Chicago. Horace Magoun (Figure 3.3; Box 7.2) had originally come

Box 7.2 Horace ("Tid") Magoun (1907–1991)

Six feet tall, well-built, and possessed of a forceful personality, Horace Magoun was an imposing presence throughout his adult life. He was born in Philadelphia; as the son of an Episcopalian minister, his upbringing was divided among a number of towns in New England as his father moved between parishes. Magoun's higher education began at Rhode Island State College and was followed by graduate studies at Syracuse University and then at Northwestern University in Chicago. There, under Walter Ranson, Magoun worked out the brain stem pathways involved in the pupillary reflexes. He also became skilled in the use of stereotaxic apparatus for localizing the stimulation of deep structures in the cat brain and of the hypothalamus in particular; he was thus able to investigate neural pathways involved in various animal behaviors. After Ranson's death, Magoun—now fully independent—continued his interest in the brain stem, this time studying the involvement of the reticular formation in cases of human bulbar poliomyelitis. In parallel animal studies he was able to show that stimulation of the reticular formation, and also of the cerebellum, altered muscle tone. However, Magoun's most important scientific work was to follow the arrival, in 1948, of Giuseppe Moruzzi as a visiting Rockefeller Fellow and their joint discovery of the reticular activating system (see main text). Published in 1949, in the first issue of *Electroencephalography and Clinical Neurophysiology,** the paper on their findings became a citation classic, and the discovery itself allegedly brought Magoun close to winning a Nobel Prize.

After this major breakthrough, Magoun made a surprising move to California, setting up a multidisciplinary Brain Research Institute in Los Angeles. In the latter part of his career he left the laboratory to devote himself to advisory work for National Institutes for Health, as well as to the organization of conferences, writing, and editing, and to a directorship for the National Research Council. His monograph, *The Waking Brain,*** published in 1958, went into a second edition. Unsurprisingly, Magoun was the recipient of many awards and other honors, including the Harvey Lectureship in 1952.

Source. Marshall, LH. Horace Winchell Magoun. June 23, 1907–March 6, 1991. *BMNAS*. 2004; 84: 251–269.

*Moruzzi G, Magoun HW. Brain stem reticular formation and activation of the EEG. *Electroencephalography and Clinical Neurophysiology*. 1949; 1: 455–473.

**Magoun, HW. *The waking brain*. Springfield, IL: Charles C. Thomas, 1952.

to Northwestern University for PhD studies under the supervision of Walter Ranson, the internationally recognized anatomist and physiologist who was also the head of the Institute of Neurology. After graduating and becoming a faculty member himself, Magoun began to study the brain stem, using a stereotactic device to guide the placement of the stimulating electrodes in different regions of the cat reticular formation. As already noted, little had been known about the function of this structure, which forms the central core of the brain stem, other than its inclusion of the respiratory centers. Indeed, in John Fulton's magisterial 675-page *Physiology of the Nervous System*, published in 1938, the reticular formation merited only eight lines, all of which had to do with respiration.

It was in 1948 that Magoun, somewhat unexpectedly, received an established neuroscientist from overseas as a temporary collaborator, Giuseppe Moruzzi (Box 7.3), who arrived from Italy as a Rockefeller Foundation Fellow. Before the outbreak of World War II Moruzzi had spent a year with Bremer in Brussels and had then joined Adrian in the latter's Cambridge laboratory. There he and his eminent colleague, already a Nobel Laureate, were the first to record the discharges of single neurons in the brain; to do so they used fine enameled silver wire electrodes inserted into the pyramidal tract—the major descending pathway from the motor area of the brain.[17] Now, 10 years later and with his war service behind him, Moruzzi was himself a professor at the University of Pisa and had become the leading neurophysiologist in Italy.

Together, Magoun and Moruzzi began to explore the possible effects on the cerebral cortex of stimulating the brain stem. Both were aware of Morison and Dempsey's demonstration of widespread cortical effects on stimulating the non-specific thalamic nuclei—effects that had nothing to do with the classical sensory pathways to the cortex. Since the reticular formation ended just below the thalamus on either side, it was logical to see if it, too, might alter cortical excitability. Using the laboratory's stereotaxic apparatus, Magoun and Moruzzi were able to stimulate small areas of the reticular formation selectively but, regardless of the site chosen, the effects on the cortex were always the same—spindles and other slow-wave EEG activity were abolished. That might have been the end of the story but for a chance increase in amplifier gain in one of the experiments. It was then apparent that, far from becoming silent, the cortex had responded to the repetitive reticular stimulation by developing low-amplitude, fast-wave EEG activity. It was exactly the kind of cortical activity that was characteristic of an animal that was alert. Magoun and Moruzzi concluded that, through its action on the excitability of the cortex, the reticular formation could control the wakefulness of the brain; further, at least part of the effect was mediated by a multisynaptic pathway running through the nonspecific nuclei of the thalamus.

The two investigators had their work on the "ascending reticular activating system" published in a new journal, and the paper became a classic citation.[18]

Box 7.3 Giuseppe Moruzzi (1910–1986)

Italy, with its many ancient universities, has long been a source of eminent scientists and scholars, many of whom had the nervous system as their main interest. Galvani discovered "animal electricity," and Golgi devised the staining method that allowed individual neurons to be visualized in all their detail. It was in this strong tradition that Giuseppe Moruzzi belonged. He was born in Parma, the son of a physician and the grandson of a professor of pathology, and it was natural that Giuseppe should himself have chosen medicine as a career. After initial training in Parma, Moruzzi—by now fascinated by the nervous system—moved to Bologna with his mentor, the neurophysiologist Mario Camis. While at a conference in Bologna honoring Galvani, Moruzzi met a number of internationally renowned neuroscientists, and this gave him the opportunity, in 1937, to work with Bremer in Brussels and then to spend the following year with Adrian. It was while he was in Cambridge that he and Adrian had been able, for the first time, to record trains of impulses

Figure B7.3. Guiseppe Moruzzi, 1965 (Photograph by Turpena; Wikimedia Commons CC SA 3.0)

arising in single cortical neurons (see main text). However, it was largely his experience with Bremer that prompted Moruzzi to make sleep mechanisms the main quest of his scientific life. Like so many other neuroscientists, however, Moruzzi's career was interrupted by World War II, during which he served as an army doctor.

Following the war, Moruzzi obtained a Rockefeller Foundation Fellowship, and this enabled him to join Magoun at Northwestern University in Chicago in 1948; it was there that the groundbreaking work on the ascending reticular activating system was carried out. By then one of the world's leading neuroscientists, Moruzzi returned to Italy as a professor at the University of Pisa; the school of neurophysiology that he created would attract many young neuroscientists and he, himself, would remain active in the laboratory for many years. Toward the end of his life Moruzzi developed parkinsonism, dying at the age of 75.

Sources. Levi-Montalcini R, Piccolino M, Wade NJ. Giuseppe Moruzzi: a tribute to a "formidable" scientist and a "formidable" man. *Brain Research Review.* 2011; 66: 256–269.
Wikipedia: Giuseppe Moruzzi.

Reading it now, there are several aspects of the publication that are of interest. One is the generosity of Magoun, the head of the laboratory, in making Moruzzi the lead author. Another is the acknowledgment of work from other laboratories—including Jasper's—which was so closely related that the discovery of the ascending reticular activating system might well have been made elsewhere. The final point, more a practical tip, was the importance of increasing the amplification ("turning up the gain") during a neurophysiological recording session. This was how Adrian had, again by chance, discovered the trains of impulses generated by sensory nerve endings, and how Paul Fatt and Bernard Katz, many years later, had found miniature synaptic potentials at the neuromuscular junction. Both discoveries were to prove of huge importance in neuroscience.[19]

After Moruzzi's return to Pisa, both he and Magoun continued to investigate the reticular formation. Moruzzi, using fine wire electrodes similar to those he had employed with Adrian many years before, was able to obtain excellent recordings from single reticular neurons and to show that the cells could be influenced by the cerebellum. Magoun, still in Chicago, demonstrated that the various sensory pathways appeared to send nerve fibers to the reticular formation while en route to the cortex; this, then, was one way in which a novel stimulus could arouse the cortex and bring it into its conscious state. It was inevitable that, because of their work on activating the cortex, both Magoun and Moruzzi should have furthered their interest in the opposite direction, that is, in the neural mechanisms underlying sleep. Armed with data and enthusiasm, the two former colleagues were to see each other again in 1953, when both took part in the Laurentian Symposium on Brain Mechanisms and Consciousness.

Among the many neuroscientists who were to follow in Magoun and Moruzzi's footsteps was one that had been on the path before—Frédéric Bremer. It was Bremer who, in 1935, had reported his innovation of the cerveau isolé preparation, the preparation in which an animal's cortex, deprived of its connections beyond the midbrain, showed the EEG features of sleep. After Magoun and Moruzzi's demonstration that the effect was due to severance of ascending fibers from the reticular activating system, Bremer himself returned to the study of sleep mechanisms. By then he was in his late 50s and still seeing patients in the neurological clinic; nevertheless, driven by his habitual curiosity, he made sure there was time for the laboratory too—indeed, Bremer would continue "hands-on" experiments well after his official retirement from the university. Among other observations Bremer showed that, far from being an independent controlling center, the reticular formation could itself be inhibited, both from the cortex and from the preoptic area of the hypothalamus. The reticular formation could also be stimulated from the cortex, and Bremer pointed out that such an effect on the brain stem had been anticipated by Adrian many years previously:

> The unfamiliar noise wakes the sleeper because the afferent message reaches the cortex and is there judged as important. The cortex signals back to the diencephalon and the rapid spread of activity ensues. The facilitation is therefore between cortex and thalamus as well as between neurone and neurone.[20]

Transmitters in the Brain

The facilitatory effect of the cortex on the brain stem was but one of several clues that there was more to the story of wakefulness and sleep. Another clue had to do with neurotransmitters, the substances released from nerve endings that were known to act on other neurons or on target tissues such as skin, heart, or gut. Thus noradrenaline was known to be released from sympathetic nerve endings while acetylcholine was not only the corresponding substance for parasympathetic nerve fibers but was also the transmitter in sympathetic ganglia and at the neuromuscular junction. Rather later, serotonin (5-hydroxy-tryptamine) was discovered in the gut and found to cause smooth muscle to contract. However, even until the early 1950s very little was known about chemical transmitters in the brain and spinal cord. Indeed, there was still the possibility that excitation from one neuron to another might be brought about by the direct flow of current generated by the nerve impulses without any transmitter substance being involved. The controversy over the nature of synaptic transmission in the central nervous system had flared up at the Cambridge meeting of the Physiological Society in 1935 in the form of a heated argument between John Eccles (for electricity) and Henry Dale (for chemicals).[21] In the following year there was a publication on the appearance of acetylcholine in slices of mammalian cortex maintained in vitro[22] and then, largely because of World War II, the field went quiet.

In 1954 a paper appeared in the *Journal of Physiology* that provided the first clear evidence that there was chemical transmission in the central nervous system, and it came from Eccles' laboratory. Eccles had made the transition from his physiology department in New Zealand to the new Australian National University in Canberra and, wasting no time, had managed to get a busy research program running in temporary accommodation. There, with Paul Fatt and Kyozo Koketsu, he succeeded in providing convincing pharmacological evidence that acetylcholine was a transmitter in the spinal cord as well as being a transmitter in the autonomic nervous system.[23] But what about transmitters in the brain?

It was not until 1963 that it finally became possible to demonstrate the liberation of acetylcholine in intact cortex; the method, pioneered by Mitchell, was to collect the transmitter in fluid-filled cylinders resting on the brain surface.[24] In the same year, and from the same institution in Cambridge, there was another paper that extended the knowledge of neurotransmitters in the brain very considerably.[25] Using methods developed in Eccles' laboratory in Canberra, Krešimir Krnjević[26] and John Phillis succeeded in making fluid-filled glass microelectrodes, each of which was surrounded by four micropipettes. Each micropipette was then filled with a solution containing a putative transmitter, small amounts of which could be released by passing an electric current through the micropipette (iontophoresis); the central microelectrode recorded impulse activity from a single neuron, thereby detecting any effect of the chemical on the excitability of the cell. Krnjević and Phillis found that, of the various substances tested, there were two that stood out—L-glutamate proved to be a powerful exciter and GABA a powerful inhibitor. And, indeed, these two compounds, with acetylcholine, have come to be accepted as the most widespread voltage-gated transmitters in the brain.

So, then, L-glutamate, GABA, acetylcholine: could there be any more transmitters in the brain, or was that the total? It was at this point that Arvid Carlsson and his colleagues in Sweden started to make their contributions to the field.

In 1956, just a few years after receiving his doctorate from Lund University, Carlsson (Figure 7.5) returned to Sweden after spending time in the United States acquiring expertise in neuropharmacology and in the use of a new kind of instrument, the spectrophotofluorimeter.[27] Together with Nils-Åke Hillarp,

Figure 7.5. Arvid Carlsson. (by Vogler; Wikimedia Commons CC BY-SA 3.0)

at the University of Lund, he was now able to show that the antipsychotic drug reserpine appeared to remove noradrenaline from the brain. At the time noradrenaline (norepinephrine) was recognized as a neurotransmitter, along with acetylcholine, but only in the autonomic nervous system. The disappearance of noradrenaline from the brains of reserpine-treated animals was associated with a striking depression of movement resembling that found in parkinsonism. Although the motor abnormality could be corrected by the administration of L-dopa, an amino acid, the clinical improvement occurred without the restoration of noradrenaline that Carlsson had anticipated. Instead, the behavioral recovery was found to be due to the return of a previously unrecognized neurotransmitter in the brain—dopamine.

It is not uncommon for a major breakthrough in science to be immediately challenged, and so it was with Carlsson and Hillarp's discovery of dopamine as a transmitter. The disbelief was all too evident at a symposium on adrenergic mechanisms that Carlsson attended in London in 1960, and the opposition was led by no lesser a person than the Nobel Laureate, Sir Henry Dale. Undeterred, Carlsson and Hillarp pressed on and, using formaldehyde, were soon able to develop an effective fluorescent method for revealing the presence of dopamine, noradrenaline, and serotonin in the brain. Of these neurotransmitters, it was noradrenaline that was released from nerve fibers that ran from the locus coeruleus, a nucleus in the brain stem reticular formation, to the nonspecific nuclei in the thalamus—the same nonspecific nuclei that had been shown to have a major role in controlling the excitability of the cortex by Morison and Dempsey a decade earlier. More than this, there were fibers from the locus coeruleus that reached the cerebral cortex directly, and these were also excitatory.[28] Here, then, was a part of the puzzle that had been missing in the study of conscious mechanisms—the noradrenergic neurons in the locus coeruleus formed the so-called reticular activating system, the lower part of the pathway that was responsible for awaking the cortex and bringing about consciousness.

Carlsson and Hillarp's work was to have another impact. Not only did they confirm that the administration of reserpine produced an animal model of parkinsonism, but they also showed that L-dopa, the precursor of dopamine, was an effective treatment. It was this treatment that would revolutionize the management of human cases of parkinsonism,[29] and it was therefore especially fitting that, in 2000, Carlsson received a Nobel Prize for the work that ultimately led to this achievement.

Hypothalamus

There was still one more piece, this time from Switzerland, that had to be fitted into the puzzle of the control of consciousness by the brain stem. Walter Hess (1881–1973) obtained his medical degree at the University of Zurich and then specialized in ophthalmology.[30] After five years of successful clinical practice, he made a mid-career change, devoting himself to neuroscience and studying the results of stimulating discrete regions on the undersurface of the cat brain and the hypothalamus in particular. Rather than recording neural activity, Hess monitored the responses of the whole animal. Depending on the site of stimulation in the hypothalamus, he was able to elicit striking autonomic effects, both parasympathetic and sympathetic, involving heart rate, blood pressure, pupil size, and posture. Moreover, these effects were usually coordinated, so that the cat might, for example, show all the behavioral features of rage. There were other experiments in which stimulation of the hypothalamus caused the very opposite, the animal going to sleep instead. It was unique and impressive work, and it was recognized by the award of a Nobel Prize to Hess in 1949.

The hypothalamus is a complex structure, however, and contains a number of functionally distinct regions. Two of these, the lateral and posterior hypothalamic areas, were later identified as the source of neurons that released a novel peptide, orexin (hypocretin), from their terminals.[31] The orexin-secreting neurons were found to project widely in the brain, providing an especially strong excitatory input to the noradrenergic cells in the locus coeruleus in the brain stem— the nucleus already identified as the key element in the reticular activating system— as well as to acetylcholine-secreting cells in the basal nucleus of the cortex. The importance of these particular pathways became evident when patients suffering from uncontrollable attacks of sleepiness (narcolepsy) were found, at autopsy, to have lacked orexin neurons in the hypothalamus. Additional studies showed that the orexin neurons were themselves liable to inhibition from another region of the hypothalamus, the preoptic nuclei. In keeping with the importance of orexin neurons in regulating sleep was their marked impulse activity during wakefulness and absent, or low, activity during sleep.

Though very much simplified, Figure 7.6 shows how the different parts of the brain interact.

Though Frédéric Bremer died at 90 in 1982, he remained mentally active in his later years and at 85 published what was then the most comprehensive account of

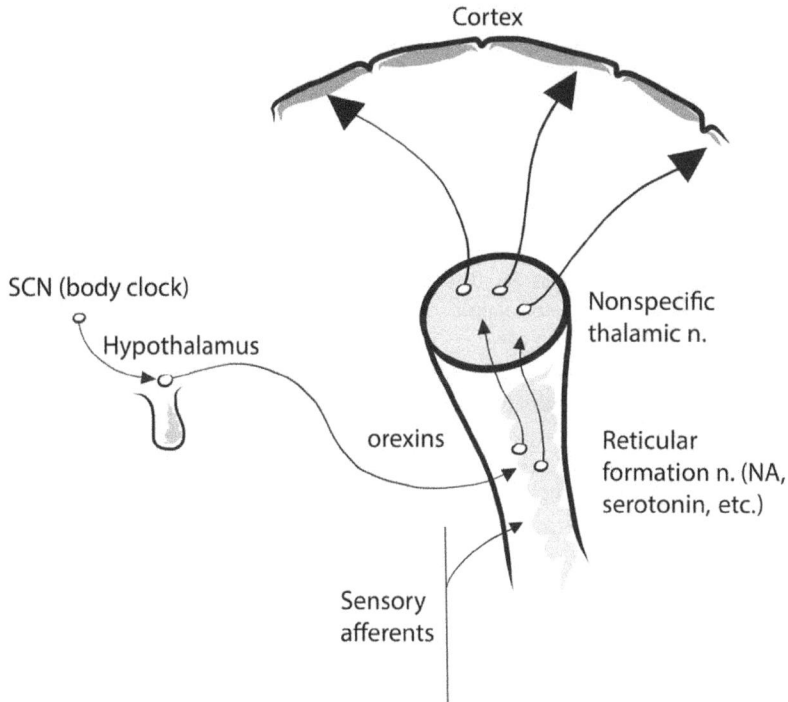

Figure 7.6. Main elements in the control of wakefulness and sleep. *SCN*, suprachiasmic nuclei; *NA*, noradrenaline.

the neural mechanisms producing sleep.[32] One suspects that, had he been alive today, he would not have been at all surprised at the wealth of discovery and detail that followed those extraordinary experiments in Brussels 80 years before. In the next chapter the nature of the electrical activity that he and others had studied in the brain is examined in more detail.

Notes and References

1. Probably the sudden, enormous barrage of impulses from the central ends of the divided sensory nerve fibers in the cervical cord would overwhelm the processing capacity of the brain and result in instantaneous loss of consciousness. However, a quadriplegic patient who has survived catastrophic injury to the high cervical cord can maintain full consciousness and a normal EEG, despite the isolation of the brain from the remainder of the nervous system.
2. See Chapter 4.
3. Bremer F. Cerveau isolé et physiologie du sommeil. *Comptes rendus des Seances de la Societe de Biologie et de ses Filiales.* 1935; 118: 1235–1241.
4. Bremer F. L'activité cérébrale au du sommeil. *Bulletin de l'Academie royale de Médicine de Belgique.* 1937; 4: 68–86.
5. Forbes' contributions to neuromuscular physiology and further details of his life are given in McComas AJ. *Galvani's spark: the story of the nerve impulse.* New York, NY: Oxford University Press, 2011.
6. This was Forbes' technician, Beningna Rempel, who married Robert Morison in the same department and was then working as Forbes' laboratory technician.
7. Forbes A, Morison BR. Cortical response to sensory stimulation under deep barbiturate narcosis. *Journal of Neurophysiology.* 1939; 2: 112–128.
8. Morison RS, Dempsey EW. A study of thalamocortical relations. *American Journal of Physiology.* 1942; 135: 281–292.
9. Dempsey EW, Morison RS. The production of rhythmically recurrent cortical potentials after localized thalamic stimulation. *American Journal of Physiology.* 1942; 135: 293–300.
10. Morison RS, Dempsey EW. Mechanisms of thalamocortical augmentation and repetition. *American Journal of Physiology.* 1942; 138: 297–308.
11. After leaving the Harvard department for the Rockefeller Foundation, Morison was first the assistant, and then the director, of Medicine and Public Health. Though no longer a neurophysiologist, he nevertheless participated fully in the 1953 Quebec symposium on Brain Mechanisms and Consciousness. Morison later re-entered academic life as a professor of biology at Cornell University and, later still, became a visiting professor at MIT. Widely respected, Morison wrote a number of books on administration, medical education, and social and ethical issues.
12. Jasper H. Diffuse projection systems: the integrative action of the thalamic reticular system. *Electroencephalography and Clinical Neurophysiology.* 1949; 1: 405–420.

13. Jasper HH. Functional properties of the thalamic reticular system. In: *Brain mechanisms and consciousness*. Delafresnaye JF, ed. Springfield, IL: Charles Thomas, 1954:373–395.
14. Li C-L, Cullen C, Jasper HH. Laminar microelectrode studies of specific somatosensory potential. *Journal of Neurophysiology*, 1956; 19: 111–130.
15. Vahe Amassian had previously carried out this kind of electrical "sink" and "source" analysis on the cortex, and the same methodology to detect sites of excitation was used elsewhere by Patrick Wall and, more extensively, John Eccles.
16. Lorente de Nó R. Cerebral cortex: architecture, intracortical connections, motor projections. In: *Physiology of the nervous system*. Fulton JF, ed. London, England: Oxford University Press, 1938:291–325.
17. Adrian ED, Moruzzi G. Rhythmic discharges from the thalamus. *Journal of Physiology*. 1939; 97: 153–199.
18. Moruzzi F, Magoun HW. Brain stem reticular formation and activation of the EEG. *Electroencephalography and Clinical Neurophysiology*. 1949; 1: 455–473.
19. The circumstances of the chance observations are described in McComas AJ. *Galvani's spark: the story of the nerve impulse*. New York, NY: Oxford University Press, 2011.
20. Adrian ED. The physiology of sleep. *Irish Medical Journal*. 1937 (June). Cited by Bremer F in *Brain mechanisms and consciousness*. Delafresnaye JF, ed. Springfield, IL: Charles Thomas, 1954:147.
21. This was the famous "Soup versus Sparks" controversy, which has been the subject of a monograph (Valenstein E. *The war of the soups and the sparks: the discovery of neurotransmitters and the dispute over how nerves communicate*. New York, NY: Columbia University Press, 2005) and a well-researched article (Marcum JA. "Soup" vs. "sparks": Alexander Forbes and the synaptic transmission controversy. *Annals of Science*. 2006; 63: 139–156).
22. Quastel JH, Tennenbaum M, Wheatley AHM. Choline ester formation in, and choline esterase activities of, tissues in vitro. *Biochemical Journal*. 1936; 30: 1668–1681.
23. Eccles' group had explored the feedback inhibition that occurs whenever a motoneuron fires an impulse; by releasing acetylcholine a branch of the motor axon excites an inhibitory interneuron (the Renshaw cell). See Eccles JC, Fatt P, Koketsu K. Cholinergic and inhibitory synapses in a pathway from motor-axon collaterals to motoneurones. *Journal of Physiology*. 1954; 126: 524–562.
24. Mitchell JF. The spontaneous and evoked release of acetylcholine from the cerebral cortex. *Journal of Physiology*. 1963; 165: 98–116.
25. Krnjević K, Phillis JW. Iontophoretic studies of neurons in the mammalian cerebral cortex. *Journal of Physiology*. 1963; 165: 274–304.
26. The young Krnjević (1927–) had to flee with his family from Yugoslavia for political reasons in 1930. and then had to flee again as World War II broke out, the boy eventually ending up in the United Kingdom after several years of schooling in South Africa. Graduating in medicine at the University of Edinburg, Krnjević then did a PhD on peripheral nerves; this was followed by postdoctoral studies, first at the University of Washington and then in Canberra with Eccles. Returning to the United Kingdom,

Krnjević carried out his work on brain neurotransmitters with Phillis in Cambridge and then accepted an offer in Canada, becoming director of the Anaesthesia Research Department and head of physiology at McGill University in Montreal. His quite extraordinary early years make for very stimulating reading, and his wide range of neuroscientific achievements, especially those involving the cerebral cortex, are impressive. See Krnjević K. The past is a foreign country. In: *The history of neuroscience in autobiography*. LR Squire, ed. New York, NY: Academic Press, 2011:7: 278–333.

27. Arvid Carlsson's life and scientific achievements are described in his 2000 Nobel lecture; the latter is accessible on the Internet through the Nobel website (www.nobelprize.org).
28. The locus coeruleus is the major noradrenergic nucleus of the brain and sends fibers not only to the cortex and thalamus but to multiple nuclei and other neural structures in the brain stem and spinal cord. Among its many functions is the control of sympathetic autonomic activity in the body. For a full account, see Samuels ER, Szabadi E. Functional neuroanatomy of the noradrenergic locus coeruleus: its roles in the regulation of arousal and autonomic function. Part I: Principles of functional organization. *Current Neuropharmacology*. 2008; 6: 235–253.
29. Nowhere was the effect of L-dopa more dramatic that in the wards of the Beth Abraham Hospital in New York's Bronx where, in 1955, there were some 80 patients with a peculiar neurological syndrome. The patients, though awake, were frozen in their postures like sufferers from parkinsonism and were often somnolent. As young men and women they had, mostly in the early 1920s, suffered an influenza-like illness associated with extreme drowsiness, a condition termed "encephalitis lethargica" by the Viennese neurologist Constantin von Economo (1876–1931). The dramatic, albeit temporary, improvement in mobility and mental attitude following the administration of L-dopa is described in Oliver Sacks' book *Awakenings* and in the subsequent film of that name starring Robin Williams as Sacks.
30. Walter Rudolph Hess' life and scientific achievements are summarized in his 1949 Nobel Lecture, accessible on the Internet through the Nobel website (www.nobelprize.org). His work on sleep is summarized in a paper at the 1953 Laurentian Symposium (Hess WR. The diencephalic sleep centre. In: *Brain mechanisms and consciousness*. Delafresnaye JF, ed. Springfield, IL: Charles Thomas, 1954:117–125).
31. Sakurai T. The neural circuit of orexin (hypcretin): maintaining sleep and wakefulness. *Nature Reviews Neuroscience*. 2007; 8: 171–181.
32. Bremer F. Cerebral hypnogenic centers. *Annals of Neurology*. 1977; 2: 1–6.

8
Electricity Works

Impulses and Slow Waves

When Descartes wrote his important *Treatise on Man*, he envisaged the brain and the nerves connected to it working by means of fluids traveling down tubular nerve fibers. That the nervous system conducts its business by electrical signals was a revelation that began with Luigi Galvani (1737–1798) in the late 18th century.[1] Working with his wife in Bologna, Galvani noticed that the calf muscles of a dissected frog twitched when there was an electrical discharge in the laboratory. Further research revealed that touching the sciatic nerve with a metal dissecting instrument would also produce a muscle twitch, and that even contact with the cut end of another nerve might be sufficient. If there was electricity in the air, as in a thunderstorm, that could also excite his nerve-muscle preparations. Not surprisingly, there was an air of magic and mystery about this early work, so much so that it provided the inspiration for Mary Shelley to create Dr. Frankenstein's Monster in her famous novel.

As described earlier, it was Emil du Bois-Reymond, in Berlin, who took the next step in unraveling the nature of the nerve impulse, by his ability to record a transient negativity in a peripheral nerve following its stimulation. Then came the important contributions of Helmholtz, Bernstein, Hodgkin, and Huxley, and finally, in our own time, Roderick Mackinnon. The nerve impulse was revealed as a reduction, even reversal, of the nerve fiber membrane potential that lasted well under a millisecond; in the thickest mammalian nerve fibers the impulses could travel at velocities of 100 ms^{-1} or even higher.

All this work had been done on peripheral nerves, however. What went on in the spinal cord and brain? In both situations the answer appeared to be slower potentials than those seen in peripheral nerves, judging by the responses that Helen Graham and Herbert Gasser were able to record from the cord, following nerve root stimulation,[2] and that Howard Bartley and George Bishop had obtained from the visual cortex in response to an optic nerve shock.[3] But even before these pioneering studies of evoked potentials there had been good evidence that slow wave potentials were important in the functioning of the brain. It was this type of activity that would have caused the galvanometer to oscillate when the English physician Richard Caton (Box 8.1), in 1875 and in front of

Box 8.1 Richard Caton (1842–1926)

As the son of a physician in Yorkshire, England, it was understandable that Richard Caton should have wished to enter his father's profession. A legacy from an aunt enabled him to study medicine in Edinburgh, and, after obtaining his degree in 1867, he moved to Liverpool, a busy port and city on the west coast of England. While practicing as a children's physician he developed an interest in research and wrote a thesis on blood cells for an MD (a higher medical degree in the United Kingdom). Two years later Caton became a lecturer in physiology in Liverpool, by which time his research interests had turned to the recording of electrical potentials in biological tissues; with the use of equipment similar to that of Du Bois Reymond in Berlin, he was able to repeat the latter's experiments on nerve and muscle. Caton then used his mirror galvanometer (which reflected a light beam onto a scale affixed to a wall) to explore the brains of animals, mostly rabbits. He found that there was a difference in potential between the surface and interior of the brain and that this potential decreased when that critical region of the brain became active—either

Figure B8.1. Richard Caton. The first physiologist to have recorded the EEG and certainly the only Lord Mayor to have done so!

following a stimulus (light or touch) or in the course of a movement (head turning or mastication). When both recording electrodes were on the surface of the brain and the animal was resting, it was possible to detect small oscillating potentials—this was the very first recording of the EEG. Caton presented his work at a meeting of the British Medical Association in Edinburgh in 1875, and a short note in the *British Medical Journal* was followed by a fuller account two years later. In recognition of his work, Caton was appointed professor of physiology in Liverpool in 1882. Later, Caton presented his work at international meetings in the United States and Russia, though by that time he had returned to clinical medicine, writing several papers on varied topics.

In addition to conducting highly original research, Caton was influential in promoting basic science teaching in medical curricula, in the development of the Medical School in Liverpool, and in the formation of the Physiological Society in the United Kingdom. He concluded a remarkable career by entering city politics, becoming Lord Mayor of Liverpool in 1907 at the age of 65; he lived a further 19 years.

---

fellow members of the British Medical Association, made his demonstration of ongoing electrical activity in the exposed brain of a rabbit.[4]

So far as is known, Caton's was the first known recording of what came to be called the EEG (electroencephalogram). Unfortunately Caton was unable to make a record of the galvanometer deflections; partly for this reason, and partly because there had not been a physiological society or a physiological journal in the United Kingdom to receive his important findings, there was little or no interest in his discovery during the next half-century. Quite independently of Caton, however, a number of investigators in Europe and Russia also began using the galvanometer to investigate the brain, again in animals. Despite their publications and correspondence, their names—Beck, Fleischl, Danilevsky, Sechenov, Tarkhanov, Cybulski, Pravdich-Neminsky—are largely forgotten today. In 1890, however, there was a dispute about priority in detecting the electrical oscillations, one that prompted Richard Caton to put his pen to paper, this time in the form of a letter in the German journal, *Centralblatt*. The letter, a polite one, drew attention to the writer's presentation to the British Medical Association 15 years earlier and finished:

> It is by no means my intention to detract from these learned physiologists, nevertheless I myself have made these observations, as described above, I have published them, so I think it must be conceded that I was already an earlier discoverer.
>
> Respectfully,
> RICHARD CATON, M.D.

Although sporadic research on the electrical oscillations continued, mostly in Russia and Poland, it seemed to attract little interest elsewhere. Then, in 1929, a publication appeared that changed everything.[5]

Berger and the EEG

The paper was in the *Archiv für Psychiatrie und Nervenkrankheiten*; surprisingly, it reported the detection of rhythmic potentials when recordings were made with electrodes applied to scalps of human subjects. At the time of these observations, Hans Berger was a physician in Jena, a medium-sized city in the eastern part of Germany that was noted both for its venerable Schiller University and for its equally famous Carl Zeiss optical works. Berger (Box 8.2) grew up in Jena, studied medicine at the university there, and was promoted to professor and head of psychiatry and neurology in 1919. Though very much an amateur in physics and electronics, his scientific obsession was to measure the energy expenditure associated with mental processes. It was, after all, his countryman, Helmholtz, who had propounded the Law of Conservation of Energy—and so Berger attempted to measure changes in cerebral blood flow and temperature, before discovering that it was also possible to record the brain's electrical activity and to influence that too by mental activity. He made his first recordings using a capillary electrometer, in which very small voltage-induced deflections of a mercury meniscus had to be magnified optically before they could be seen or recorded. Although Adrian, in Cambridge, had made his groundbreaking discovery of impulse trains in sensory nerve fibers with a capillary electrometer, it was a very capricious device, and Berger replaced it first with a string galvanometer and then with a Siemens double-coil galvanometer, as soon as these superior instruments became available. For recording electrodes Berger used needles inserted into the scalp or padded pieces of silver foil applied to the surface of the head and held in place with a rubber bandage. Having started these experiments in 1924, five years elapsed before Berger was ready to put any of his observations in print and a further five years before his publications had been completed.

Berger's key observation was that there were oscillating potentials, with a frequency of around 10 Hz, that could be detected over the head. The rhythm was apparent when the subject was relaxed with the eyes closed but disappeared when the eyes were opened or when the subject's attention was otherwise engaged, as in performing mental arithmetic. Evidence that the potentials were generated by the brain came from the observation that the potentials were larger when a recording electrode could be placed over an opening in the skull (in a neurosurgical patient).

Box 8.2 Hans Berger (1873–1941)

Hans Berger was born in Neuses, Germany, and, after completing high school, attended the Friedrich Schiller University in Jena to study mathematics; after one semester, however, he left university for military service in the cavalry. It was then that he had a near-fatal accident that coincided with an intuition on the part of his distant sister, and this combination led to Berger's belief in telepathy. Returning to Jena, this time to study medicine, Berger's interest in telepathy persisted and led him to choose psychiatry as his specialty. His research interest was in discovering evidence of physical changes in the brain that were associated with mental events, and his first attempts at recording electrical activity in animal brains began in 1902. A rigid and secretive man, Berger carried out his experiments in a small building in the grounds of the clinic, and, though the results were carefully noted at the time, he told no one about his work. Promoted to senior lecturer in psychiatry in 1906, Berger was subsequently made professor of psychiatry and neurology in succession to Otto Binswanger in 1919. Before that, however, he had married into the German nobility and served in World War I as a psychiatrist on the Western Front. After the war Berger returned to Jena and his unusual research, extending his studies to the human brain. It was in 1925, using a Siemens double-coil galvanometer, that Berger succeeded, with scalp electrodes, in making the first recordings of spontaneous electrical activity in the human brain (see main text); this activity he termed the *Elektrenkephalogramm*—in English, the

Figure B8.2. Hans Berger, possibly in 1920. (Wikimedia Commons)

"electroencephalogram" (EEG). Ever secretive, and perhaps uncertain about the basis of his discovery, Berger delayed publication for four years. It was only after the verification of his findings by Adrian and Matthews, however, that the importance of Berger's discovery was universally recognized and that he achieved fame. In 1935 Grey Walter (see Box 14.2) left Cambridge to visit Berger in Jena and found him a "modest and dignified person, full of good humour" (though sadly lacking in expertise in electronics).

These were to be Berger's best years. Obliged to retire at the age of 65 in 1938, Berger became depressed, admitted himself to hospital, and hanged himself there on June 1, 1941.

Berger's publications came at a time when German neuroscience was still recovering from the upheaval and loss of life engendered by World War I. It was also the time of the ascendancy of Hitler and the National Socialist Party. With the subsequent decree that forbade Jews from holding professional positions in universities, government, and hospitals, the exodus would begin—to the enormous benefit of the countries accepting the talented immigrants. But despite the loss of Jewish colleagues there were still a number of eminent German neuroscientists. Among these, Oskar Vogt (1870–1955) was studying the fine architecture of the cerebral cortex in his neurological institute in Berlin while Paul Hoffmann (1884–1962), in Freiburg, was employing his eponymous reflex in the investigation of spinal cord function. On the neurosurgical side, Otfrid Foerster (1873–1941), in Breslau, was adding to knowledge of human neuroanatomy and neurophysiology through his neurosurgical innovations. It is likely that German neuroscientists of their quality and prestige would have regarded Berger's work with suspicion; even if his findings were true, the work would have been viewed as "fringe" in contrast to their own mainstream activities. Was it not a quixotic undertaking to attempt to measure the energy expended in thinking?

Verification in Cambridge

In Britain, however, Berger's publications provoked interest, even if his findings and their interpretation seemed unlikely. In 1934 Edgar Adrian (Box 8.3) was at the pinnacle of his career. The son of a barrister, Adrian studied physiology at Cambridge and, in the years before World War 1, undertook research on the nerve impulse with his mentor, Keith Lucas (1879–1916). Following Lucas' death in a flying accident during the war, Adrian turned his attention to sensory mechanisms and—initially through a chance observation—discovered that there were sensory receptors in muscle and skin that fired trains of nerve impulses throughout an applied stimulus. Like Lucas before him, Adrian had

Box 8.3 Edgar, Lord Adrian (1889–1977)

Edgar Adrian was born in London, the son of a prominent barrister. An exceptional scholar, Adrian was educated at Westminster School before proceeding to Cambridge to study for the Natural Sciences Tripos. After exposure to physiology, he became extremely interested in that subject and began work in the Physiological Laboratory under the mentorship of Keith Lucas. At a time when physiology, at least in the United Kingdom, was still in its infancy, Lucas became the nation's foremost neurophysiologist, his main preoccupation being the nature of the nerve impulse. Together Adrian and Lucas were able to show that the impulse was an all-or-none phenomenon in the sense that, once elicited, it assumed its full amplitude as it traveled along the nerve fiber; further, the energy for the transmission of the impulse resided in the nerve fiber. They drew an analogy with a train of gunpowder—once ignited with a match, the flame spread along the train as the powder was consumed. After Lucas' premature death through a flying accident in World War I, Adrian was obliged to pursue neurophysiology at Cambridge by himself and decided to leave the mechanism of the nerve impulse for a study of impulse generation and propagation from sensory receptors. This work, for which he would ultimately win a Nobel Prize, began with the chance observation of nerve activity when a muscle was hanging down and stretched by its own weight. Having made his original observations with a capillary electrometer, Adrian obtained superior recordings by adopting valve amplification and a Matthews oscillograph. With a remarkable gift for reaching penetrating conclusions from apparently simple experiments, Adrian turned his attention from one part of the nervous system to another, usually preferring to work alone. The author of several monographs and many papers, Adrian was able to express his findings and conclusions in beautiful prose. The photographs of the elderly Adrian, doyen of British neurophysiologists, are misleading, however—this was a man who had lived dangerously (climbing college roofs at night), had enjoyed mountaineering, had been an expert fencer, and had played a memorable practical joke in his youth, exhibiting his own hasty paintings as those of a master. A Nobel Laureate in 1932, with Charles Sherrington, Adrian was given a baronetcy in 1955 in recognition of his research into the nervous system; at that time he was both Master of Trinity College, Cambridge, and president of the Royal Society.

Source. McComas AJ. *Galvani's spark: the story of the nerve impulse*. New York, NY: Oxford University Press, 2011. For a photograph of Adrian, see Figure 7.2.

imagination and scientific daring and was prepared to turn his attention to any problem that interested him, usually developing novel preparations to do so. In 1932, his research on sensation was acknowledged by the award of a Nobel Prize, an honor that was shared with his Oxford counterpart, Charles Sherrington.

Also present in Adrian's Cambridge laboratory in 1934 was a younger man similarly gifted. Bryan Matthews (1906–1986) was not only a fine neurophysiologist but an inventor—he had been the first to come up with the idea of "push-pull" amplification as a means of improving signal detection from noise. He had also designed and built a moving–iron oscillograph which, although not having the frequency response of a cathode ray oscilloscope, was nevertheless capable of producing excellent recordings of impulse trains as well as of slower events. Together, Adrian and his junior colleague set out to explore Berger's claims. Referring to Berger's report of a brain rhythm, they began by expressing their skepticism: "We found it difficult to accept the view that such uniform activity could occur throughout the brain in a conscious subject, and as this seemed to us to be Berger's conclusion we decided to repeat his experiments."[6]

Using another Matthew's invention, the ink-writing oscillograph, and with padded copper gauze electrodes on Adrian's scalp, they found that there were indeed rhythmic potentials when Adrian was relaxed with his eyes closed (Figure 8.1). With the eyes open, any attempt to see a visual pattern would abolish the rhythm, as would other forms of sensory stimulation and mental activity. Rather than the rhythm being a feature of the entire cortex, as Berger had thought, Adrian and Matthews were able to show that it was generated in

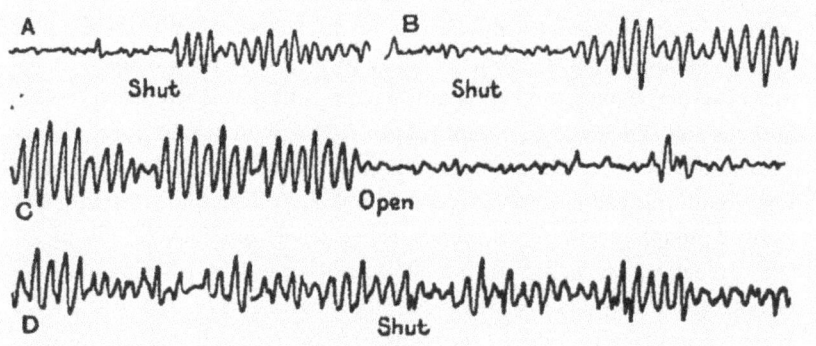

Figure 8.1. EEG recordings by Adrian and Matthews. 10–12 Hz activity appears after the eyes have been closed in Adrian (*A*) and Matthews (*B*) and is blocked in Adrian on opening the eyes (*C*). In a dark environment, the activity continues regardless of whether the eyes are open or closed (*D*). (From: Adrian ED, Matthews BHC. The Berger rhythm: potential changes from the occipital lobes in Man. *Brain*. 1934; 57: 355–383)

the visual cortical areas at the back of the brain. In the plain simple language so characteristic of Adrian, "In man a large area is normally occupied with visual activities; thus when the area has nothing to do and is free to develop a synchronous beat the potential changes are large enough to be detected outside the skull."

Adrian and Matthews had one more act before they were to leave the "Berger rhythm." At the Cambridge meeting of the Physiological Society in the summer of 1934 and in front of their fellow physiologists, they put on a demonstration, with Adrian the subject and Matthews the experimenter. With Adrian's beautiful 10 Hz rhythm present for all to see, there could no longer be any doubt as to the existence of EEG activity. An entirely new field of research had opened up, a field in which both Grey Walter[7] and Herbert Jasper,[8] among others, would play prominent roles. But what sort of neural activity was responsible for slow waves?

Slow Waves as Synaptic Potentials

Hitherto much of the neurophysiological research had been devoted to understanding the ionic mechanisms responsible for the nerve impulse, and it was therefore natural to wonder if the slow waves were simply summated impulses. This possibility looked unlikely from the experiments of Gasser and Graham, previously mentioned; in their recordings from the spinal cord it had been easy to distinguish the brief action potential volley from the slower waves that followed it. The work on the visual pathway coming from George Bishop's laboratory at Washington University in St. Louis also had to be considered. Bishop (Figure 8.2) played a major role in studying the nerve action potential by the use of the cathode ray oscilloscope but, because of a publication, had then been banished from the Department of Physiology by its head, Joseph Erlanger.[9] Having found academic refuge in the pharmacology department, the resourceful Bishop began to investigate the visual pathway. With the rabbit as their preparation, he and Howard Bartley discovered that electrical stimulation of the optic nerve produced slow wave responses in the visual cortex. Addressing the possibility that the potentials were summated action potentials in bundles of cortical nerve fibers, Bishop and Bartley performed a simple but direct experiment—they laid an excised vagus nerve along a slit in the rabbit cortex and found that stimulation of the nerve yielded only very small potentials, the small size due to the shunting effect of the surrounding cortex.[10]

With nerve fiber action potentials excluded as a cause, one clue to the possible nature of the slow potentials came from outside the central nervous system. Stimulation of a muscle through its nerve produced an end-plate potential lasting 20 ms or so, and even longer potentials developed in the sympathetic

Figure 8.2. George Bishop. A gifted and versatile neurophysiologist, he was unlucky not to have shared the Nobel Prize with Erlanger and Gasser in 1944. (Reproduced courtesy of the Becker Medical Library, Washington University School of Medicine)

ganglion following nerve stimulation. Both observations raised the possibility that the slow waves in the brain were also synaptic potentials. More information about synaptic potentials in neurons came from John Eccles,[11] who had worked on both muscle and sympathetic ganglia during his time in Oxford and in Sydney, Australia. Later, at the University of Otago in New Zealand, he followed up his doctoral studies on the spinal cord, studies that he had carried out with Sherrington in Oxford, by investigating synaptic potentials with intracellular recordings from motoneurons.

The new work was highly original, and the fact that it was carried out in a little-known university on the far side of the world made it all the more remarkable. It was not a one-person show, however, for Eccles had been fortunate in recruiting the services of two gifted junior colleagues: Lawrence Brock was a highly capable and versatile experimenter who would later become a surgeon and then a neuroradiologist; Jack Coombs was a shy but brilliant physicist who was able to design and build the essential stimulating and recording systems.

In planning the new enterprise, Eccles was very much aware of the success that had come from intracellular recordings in the squid giant axon by the Cambridge physiologists Alan Hodgkin and Andrew Huxley in their experiments at the Marine Biological Laboratory in Plymouth. Eccles also knew that glass capillary microelectrodes had been used to penetrate and record from single muscle fibers by Ralph Gerard and Judith Graham in Chicago and by Alan Hodgkin a little later.[12] The electrodes were easily made, though skill was required to obtain the ideal taper and tip. The middle section of a piece of capillary glass tubing was heated and the two ends of the tubing rapidly pulled apart; though the tip of a broken end would be 1μm or less in diameter, it could still be filled with an electrolyte solution. With Brock's expertise ensuring a continuing supply of quality electrodes, Eccles proceeded to make intracellular recordings from cat motoneurons.[13] He found that stimulation of an excitatory nerve could reduce the membrane potential of the motoneuron for 10 ms or more ("depolarization"), while an inhibitory nerve had the reverse effect, raising the membrane potential for a similar period ("hyperpolarization"). Eccles termed the changes in potential the excitatory postsynaptic potentials (EPSPs) and inhibitory postsynaptic potentials (IPSPs), respectively Figure 8.3).

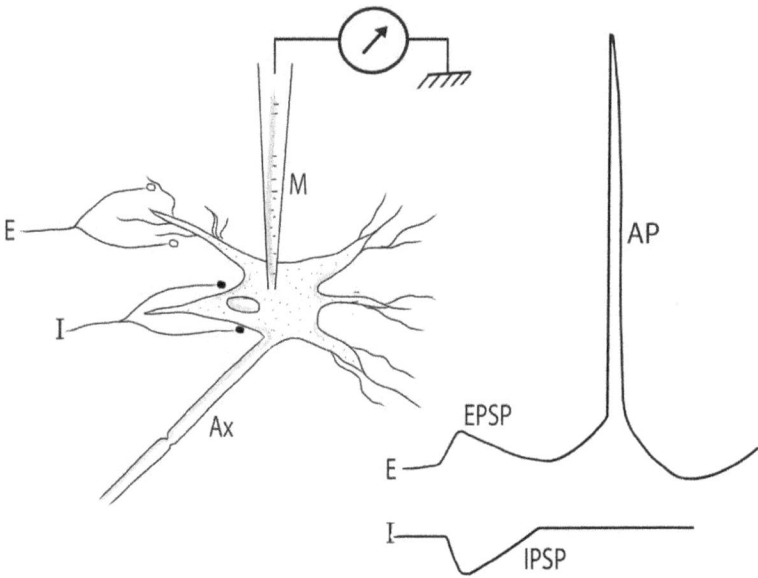

Figure 8.3. Simulated intracellular recording from a motoneuron with a microelectrode (*M*); stimulation of an excitatory nerve (*E*) elicits an EPSP (excitatory postsynaptic potential—a depolarization of the motoneuron membrane) while stimulation of an inhibitory nerve (*I*) produces an IPSP (inhibitory postsynaptic potential, a hyperpolarization). A sufficiently large EPSP triggers an action potential (AP).

After publication of this seminal work, Eccles moved to the new Australian National University in Canberra, taking Coombs with him. It was there that he began a golden period of research on the brain and spinal cord, attracting senior neuroscientists from around the world as temporary collaborators. So great was the impact of the initial spinal cord work, however, that it tended to overshadow the achievement of two other neurophysiologists. In the Physiology Department of the University of Washington, Walter Woodbury and Harry Patton also succeeded in making intracellular recordings in the cat spinal cord at about the same time as Eccles and were the first to record from spinal interneurons. For various reasons, however, the two Americans did not pursue their initial success and it was left to others, including Eccles, to show that EPSPs and IPSPs could be evoked in cortical neurons and, indeed, throughout the central nervous system; importantly, the postsynaptic potentials were responsible for some of the cortical oscillations, including Berger's 10 Hz rhythm.

The next question concerned the origin of the brain rhythms. Did they arise spontaneously in the cortex, or were the cortical cells being driven from elsewhere? The question arose because of the findings of Morison and Dempsey. These investigators, it will be recalled,[14] showed that cells in the nonspecific thalamic nuclei projected to wide areas of the cortex and, when stimulated, could initiate rhythmical activity in the cortex. Further, the thalamic cells themselves exhibited rhythmic activity, and Thomas Sears and Per Andersen would later suggest how, through a combination of nerve fiber branching and IPSPs, this activity could build up within a thalamic nucleus.[15] On the other hand, two neurophysiologists in Montreal, Kristiansen and Courtois,[16] had shown that, even if an area of cortex had been isolated by cutting all the connections from the rest of the brain, it was still capable of generating spontaneous oscillatory activity. Further, it had long been known, though histological studies, that there was a strong projection of nerve fibers from all the cortical regions back to the thalamus; it was subsequently shown that, by means of these fibers, electrical stimulation of a cortical area could elicit rhythmical potentials in the thalamus. It all seemed very complicated, but as the experimental results accumulated over several decades, one fact became clear—rather than considering the workings of the cortex and thalamus separately, or even suggesting that one was more important for consciousness than the other, the two structures had to be considered as a functional unit; they always worked in tandem, with each influencing the other.

Single Cell Studies in Cortex

To summarize, slow waves, whether generated spontaneously or in response to excitation elsewhere, were shown to be a prominent electrical sign of activity in the brain. But the slow waves, at least the excitatory ones (EPSPs), only had consequences if they were sufficiently large to set off the other type of electrical activity in the nervous system, the action potential (nerve impulse). As already noted, the action potential had been very well studied over many years using peripheral nerve fibers, some of them very large, with a variety of electrical, chemical, and ultrastructural techniques. It proved much more difficult to obtain similarly detailed information about the actions potentials of neurons in the brain by intracellular recordings, though in 1956 Charles Phillips, in Oxford, reported doing this in the cat motor cortex[17] and, in the previous year, Denise Albe-Fessard and Pierre Buser (in Paris) had a note published on their own intracellular results in the same motor area.[18]

The most important deterrent to successful recording was continuous movement of the brain tissue from arterial pulsation and respiration; even if one was lucky enough to insert a capillary electrode into a neuron, the glass tip was almost certain to break. One attempted remedy, pioneered by Edward Evarts at the National Institutes of Health in Bethesda, Maryland, was to place the electrode inside a sealed metal chamber that had been attached to the skull surrounding the exposed brain;[19] however, the most satisfactory solution to the pulsation problem came, quite unexpectedly, from biochemistry. Some years previously Hugh McIlwain, while still a PhD student in Newcastle upon Tyne in the United Kingdom, had studied energy expenditure during brain activity by carrying out experiments on slices of brain contained in a bath of nutrient fluid; he continued these studies in London.[20] Despite initial skepticism from neurophysiologists, it was confirmed that intact neurons inside the slices remained in good condition, as judged by their resting membrane potentials.[21] As there was no pulsation, stable intracellular recordings enabled all the electrical properties of a neuron to the studied—together with the underlying ionic permeabilities (conductances) of its membrane. Not only that, but the morphology of the neuron could be demonstrated by injection of a dye through the capillary electrode, as could the connections between that neuron and others.

Unfortunately for the neurophysiologists, the experiments on slices revealed even greater complexity in the cortex. For example, whereas a peripheral nerve fiber depended on two major conductances, those for sodium and potassium, for the initiation and propagation of the impulse, a single cortical neuron contained a host of conductances. This finding, allied to the branching patterns of the dendrites, suggested to some that each neuron might be acting as a computer. Thus a human cortex containing an average of 28×10^9 neurons would contain the same number of computers! Nor did the recordings of impulse activity in the slices provide the answers, for it appeared that there were different types of neuron—some fired slow trains of single impulses, others fired in bursts of impulses, and yet others discharged at very high rates (up to 400 impulses a second). Also, in addition to the 10 Hz cortical rhythm that Berger had discovered, there were slower and faster rhythms as well. And then one had to deal with the fact that, in addition to the pyramidal neurons, which formed the majority of cortical neurons, there were also the inhibitory cells (the "basket" and "chandelier" cells) to consider. As if there were not enough problems, lurking in the background was Adrian and Matthews' old observation that the optic ganglion of the lowly water beetle also generated a rhythm indistinguishable from that of the human EEG and this rhythm, too, was abolished by light![6] Finally, there was the role of the thalamus to be considered; the thalamic nuclei also exhibited rhythmic activity and their axons projected to the cortex. The situation was one that could only be resolved by extraordinary skill and dedication.

Interactions Between Thalamus and Cortex

Among the many who attempted to resolve the complexities of the relationship between cortex and thalamus were Dominick Purpura and his colleagues at Columbia University in New York who, in 1964, appear to have been the first to make intracellular recordings from cortical and thalamic neurons during rhythmic activity.[22] Rodolfo Llinás, too, became a major figure in the field, combining in vitro, animal, and even human studies.[23] But the person who devoted almost his entire scientific life to the thalamocortical relationship, and to the nature of the rhythmic activity, was Mircea Steriade (Box 8.4). Though born and educated in Romania, Steriade spent the bulk of his career in the Université de Laval in Quebec. With a combination of strategies—holding the head steady in a stereotaxic frame, cementing observation chambers to the skull, and employing all the little tricks that come with experience—Steriade was able to obtain stable recordings for an hour or more, with the tip of a glass capillary electrode remaining inside a cortical or a thalamic neuron. In this way he was able to examine the behavior of the same neuron when the animal drifted from alertness into sleep, or when it undertook a "voluntary" movement. In addition to observing the natural firing pattern, Steriade could see how a cell responded to the injection of current through the electrode.

What Steriade found was very important. First, it was a mistake to classify cells on the basis of their impulse firing pattern; all cells—whether pyramidal neurons or inhibitory neurons, whether in the superficial or in the deeper layers of the cortex—could exhibit identical impulse firing patterns. Second, the firing pattern would change as the animal moved between sleeping and waking states, with the very rapid discharges occurring in the latter. Steriade also found that, although the classic 10 Hz rhythm was generated in the thalamic nuclei, in keeping with the old results of Morison and Dempsey, it could be prompted by a

Box 8.4 Mircea Steriade (1924–2006)

Mircea Steriade was born in Bucharest, Romania, in the summer of 1924. Under his mother's strong influence, Steriade studied hard at high school despite the distractions and difficulties of World War II. He enrolled in the medical school soon after the end of hostilities, graduating in 1952. Having developed an interest in neurophysiology in his medical studies, and particularly in the relationship between the cortex and deeper structures, he then embarked on a doctorate in science. During this time he combined clinical work in neurology with the investigation of interactions between cortex and cerebellum in animals. His thesis was published in France and drew praise from Penfield and Brodal, as well as from Frédéric Bremer. Aware of the latter's prewar work in developing the encéphale isolé and cerveau isolé preparations, Steriade then spent a year working with Bremer in Brussels as a postdoctoral fellow. On returning to Romania, Steriade was appointed head of the Neurophysiology Laboratory at the Institute of Neurology in Bucharest.

Figure B8.4. Mircea Steriade. A master of intracellular recordings in vivo. (Photograph courtesy of Dr Igor Timofeev)

After 10 years in this position, coping with the difficulties imposed by a communist government, he made his final move—a rather surprising one—to the Université Laval in Quebec City, Canada. There, as professor of physiology and the university's first neurophysiologist, he was to remain for the next 38 years, building a leading international laboratory—in a single room—from scratch. His theme, and that of the many graduate students that came to work with him, was the electrophysiological investigation of thalamo-cortical interrelationships. A second interest, maintained over many years, was the neuronal basis of epilepsy. Dedicated to his science, Steriade would arrive at his office at 6:00 AM and, after five hours of concentrated work, would break off to spend the best part of an hour in the swimming pool, before having lunch and resuming activity in the laboratory and office; it was a regimen that he would maintain to the end.

Sources. Steriade M. *The history of neuroscience in autobiography.* Squire LR, ed. New York, NY: Academic Press, 2004:4:486–519.
Timofeev I. Mircea Steriade (1924–2006). *Neuroscience.* 2006; 142: 917–920.

discharge to the thalamus from the cortex. Further, the cortex was, by itself, capable of producing all the other kinds of rhythms, slow and fast, as well as even slower cortical oscillations. Steriade was able to summarize his work, and that of other cortical neurophysiologists, in an important review that came out within a year of his death.[24]

Though the human brain had many billions of cortical neurons it was now possible to recognize patterns of activity, patterns that would be relevant to the study of consciousness. One such pattern, a marked increase in firing rate on presentation of an appropriate stimulus, led to the discoveries of a very important principle in the processing of information and of a very special type of neuron. It was a neuron whose existence had been posited by some and denied by others, and it had been given an intriguing name: the "grandmother" cell.

Notes and References

1. The history of research into the nature of the nerve impulse, from Galvani to MacKinnon, has been described in McComas AJ. *Galvani's spark: the story of the nerve impulse.* New York, NY: Oxford University Press, 2011.
2. Gasser HS, Graham HT. Potentials produced in the spinal cord by stimulation of dorsal roots. *American Journal of Physiology.* 1933; 103: 303–320.
3. Bartley SH, Bishop GH. The cortical response to stimulation of the optic nerve in the rabbit. *American Journal of Physiology.* 1932; 103: 159–172.

4. The history of the EEG is described fully and beautifully by Mary Brazier in her monograph: *A History of the Electrical Activity of the Brain* (London, England: Pitman Medical, 1961). This delightful little book contains photographs of all those involved in the early studies of the electrical activity of the brain up to, and including, Berger.
5. Berger H. Über das Elektrenkephalogramm des Menschen. *Archiv für Psychiatrie und Nervenkrankheiten*. 1929; 87: 527–570.
6. Adrian ED, Matthews BHC. The Berger rhythm: potential changes from the occipital lobes in man. *Brain*. 1934; 57: 355–383.
7. For further information on Grey Walter, see Box 14.2
8. For further information on Herbert Jasper, see Box 3.4.
9. The circumstances that led to Bishop's banishment from the Physiology Department are that he had succeeded in demonstrating impulse conduction in very fine unmyelinated nerve fibers for the first time—something that Erlanger, as a senior scientist and departmental head, should have done earlier or, in Erlanger's opinion, at least been consulted about.
10. Bartley SH, Bishop GH. Factors determining the form of the electric response from the optic cortex of the rabbit. *American Journal of Physiology*. 1932; 103: 172–184.
11. For more information on Eccles, see Box 3.3.
12. Well before Graham and Gerard's studies, fluid-filled glass microelectrodes had been used for recording from the cells of algae, but these authors were the first to show that such electrodes could also be employed for intracellular recordings in mammalian cells (in their case, from muscle fibers). See Bretag AH. The glass micropipette electrode: a history of its inventors and users to 1950. *Journal of General Physiology*. 2017; 149(4): 417–430.
13. Eccles' early studies with intracellular recordings in spinal motoneurons are summarized in his monograph, *The Physiology of Nerve Cells* (Baltimore, MD: Johns Hopkins Press, 1957). A very comprehensive history of the early intracellular work is Stuart DG, Brownstone RM. The beginning of intracellular recording in spinal neurons: facts, reflections and speculations. *Brain Research*. 2011; 1409: 62–92 (available at www.sciencedirect.com).
14. See Chapter 7.
15. Andersen P, Sears TA. The role of inhibition in the phasing of spontaneous thalamocortical discharges. *Journal of Physiology*. 1964; 173: 459–480. (The work was done while the two authors were visiting scientists with John Eccles in Canberra. Allegedly, it was Andersen who was largely responsible for Eccles' decision to move from the spinal cord and work on the thalamus.)
16. Kristiansen K, Courtois G. Rhythmic electrical activity from isolated cerebral cortex. *Electroencephalography and Clinical Neurophysiology*. 1949; 1: 265–272.
17. Phillips CG. Intracellular records from Betz cells in the cat. *Quarterly Journal of Experimental Physiology*. 1956; 41: 58–69. For a photograph of Phillips, see Figure B6.2.
18. Albe-Fessard D, Buser P. Activités intracellulaires recueillies dans le cortex sigmoide du chat: participation des neurons pyramidaux au " potential évoque" somesthésique. *Journal de Physiologie*. 1955; 47: 67–69.

19. Evarts EV. A technique for recording activity of subcortical neurons in moving animals. *Electroencephalography and Clinical Neurophysiology.* 1968; 24: 83–86.
20. McIlwain H, Buchel L, Cheshire JX. The inorganic phosphate and phosphocreatine of brain especially during metabolism in vitro. *Biochemistry Journal.* 1951; 48: 12–20.
21. Hillman HH, McIlwain H. Membrane potentials in mammalian cerebral tissue in vitro; dependence on ionic environment. *Journal of Physiology.* 1961; 157: 263–278.
22. Purpura DP, Shoffer RJ. Cortical intracellular potentials during augmenting and recruiting responses. I. Effects of injected hyperpolarizing currents on evoked membrane potential changes. *Journal of Neurophysiology.* 1964; 27(2): 133–151.
23. For information on Rodolfo Llinás see Box 12.1.
24. Steriade M. Neocortical cell classes are flexible entities. *Nature Reviews Neuroscience.* 2004; 5: 121–134.

9
Single Units and Grandmother Cells

Single Neuron Recordings

The novel findings made with scalp electrode recordings by Berger and then by Adrian and Matthews were described in the previous chapter. Useful as such studies were, they had the limitation that the potentials were the sum of activity in thousands of cortical neurons within range of the pick-up electrodes. It was the activities of individual cells that had to be known if the workings of the brain were to be fully understood, and this was especially true for consciousness. The situation was far easier for those for those who were experimenting at the level of the spinal cord or brain stem and who were interested in movement. There it was possible not only to record the discharges of individual motoneurons but, as was shown many years ago, to estimate the number of motoneurons for any given human muscle.[1] The ease of recording motoneuron activity was possible because the several hundred muscle fibers supplied by a single motoneuron (a "motor unit") acted as biological amplifiers. Nor was it obligatory to insert a needle into the muscle for the recording—a surface electrode was often sufficient for the task at hand.

In animal studies those central nervous system physiologists working in the mid-20th century tackled the problem of neuron recording by using electrodes that had tips smaller than the cell bodies. As described in the previous chapter, the first such microelectrodes to be widely used were made from capillary glass tubing, and it was this type of electrode that had been employed in Ralph Gerard's laboratory in Chicago for measuring the resting membrane potentials of muscle fibers. The glass microelectrodes did have one disadvantage, however: the tendency of the tips to break, so that prolonged recordings from the same neuron were difficult. While this was a nuisance in animal experiments, in human subjects the breakages prevented the use of the electrodes—as in patients being evaluated for deep brain neurosurgery, for example.

One solution to the problem came from David Hubel, then investigating cells in the mammalian visual cortex while working at the Johns Hopkins Hospital in Baltimore, Maryland. Hubel had taken straight lengths of tungsten wire and etched the ends by repeatedly dipping them in acid while passing current through them. Using varnish, each piece of tungsten was then insulated almost to the tip of the tapered end. Because they were made of tungsten, the electrodes did not bend or break as they penetrated the brain, and they were found to be excellent for recording impulses in single or small groups of neurons.

The first patients to be investigated with such electrodes were those with parkinsonism. Following a chance observation, it had been found that the coarse parkinsonian tremor could be abolished by making a small lesion in the thalamus. The problem was that the thalamus lay deeply within the brain and, since human heads and brains varied considerably in size and shape, there was no guarantee that an inserted probe would reach the correct target, even if the best brain atlas were consulted. However, in a small number of neurosurgical centers the neurophysiologists, by recording with long Hubel-type electrodes, were able to identify the somatosensory nuclei within the thalamus and thereby direct the neurosurgeon's probe to the target nucleus immediately anterior. In such sessions neurophysiologists were able to record impulses fired by thalamic neurons responsive to touching small areas of skin, usually on the face or hand (Figure 9.1); other cells only fired when a joint was moved.

Useful though the thalamic recordings were, they added little to what had already been learned about somatosensory pathways from animal studies. Not

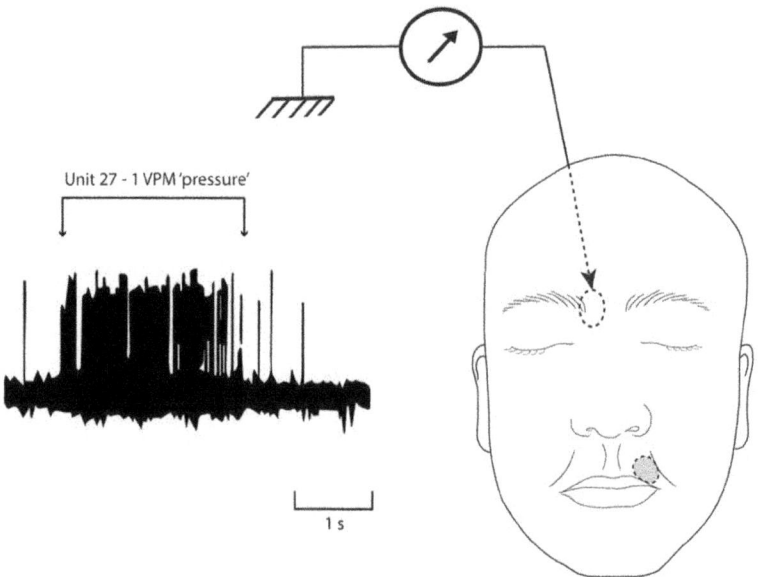

Figure 9.1. Tungsten electrode recording from a neuron in the human thalamic VPM nucleus that responded to pressure above the lip on the opposite side. (Author's recording)

only that, but the recordings presented an enigma, for many neurons were encountered by the electrodes in other parts of the thalamus that failed to show any alteration in their baseline activity no matter what testing strategy was employed. Given these rather disappointing results from the thalamus, it seemed possible that single cell recordings from the human cortex—if they were ever made—might prove similarly unproductive. Instead, it was just this type of study that helped to resolve a fundamental question concerning the way the brain processed sensory information. Were the various features of a stimulus simultaneously processed in different cortical areas and the results then combined to yield the perception? Or did the brain create and store templates for each perception, to which it could refer in future? Could there be such a thing as a "grandmother cell"?

The Grandmother Cell

Jerry Lettvin (Box 9.1) was responsible for the term. A neurophysiologist, Lettvin had spent much of his academic life investigating visual mechanisms in the frog's retina at the Massachussetts Institute of Technology in Boston. The "mother" (later "grandmother") term appeared in a fictitious (and very amusing) story that he told his students and that he also recounted in a letter. In this story a neurosurgeon is called upon to treat a patient who has a pathological dislike of his mother. The neurosurgeon, a Russian, deals with the problem by ablating all those nerve cells in the patient's brain in which the mother was represented. Lettvin chose Philip Roth's character, Portnoy, (the principal character in *Portnoy's Complaint*) as the patient (Box 9.2).

Why did Lettvin invent such an imaginative story? The answer lies in large part with Horace Barlow, the recipient of Lettvin's letter. Barlow, a Cambridge physiologist and a great-grandson of Charles Darwin, had himself explored the frog retina[2] and given considerable thought as to how information might be processed by the visual system not only in the frog but in mammals too. Largely on theoretical grounds, he thought it likely that, as the information was passed on from one set of neurons to another along a visual pathway, so fewer and fewer cells would be involved—moreover, these cells would become more and more demanding in what could excite them.[3] Lettvin had already seen a clear indication of such a process in his vision experiments. Thus the retina in the frog, as in mammals, is a surprisingly complex structure for, in addition to the rods and cones, which act as receptors for light, there are amacrine and horizontal cells and, finally, the ganglion cells that give rise to the optic nerve fibers. The ganglion cells, Lettvin found, responded best to small dark shapes moved across the visual field—the sort of shapes that can cause normal frogs to jump at them and protrude their

Box 9.1 Jerome (Jerry) Lettvin (1920–2011)

Jerry Lettvin once described himself as "an overweight slob, disheveled and careless of appearance." In the same autobiographical essay he insisted that he had never been a true scientist and that he was totally incompetent in the laboratory!

What kind of man was this, then, who could write so frankly and dismissively of himself? The answer, surely, is that Lettvin was one of the most colorful, imaginative, and stimulating neuroscientists of his generation, and one possessed of a great sense of humor.

Lettvin was born in Chicago to Ukrainian Jewish parents who had emigrated to the United States just prior to World War I. He had attended the University of Chicago and caved in to his mother's demands to study medicine—left to himself, he would gladly have become a poet and writer. After graduating from the University of Illinois in 1943, Lettvin served

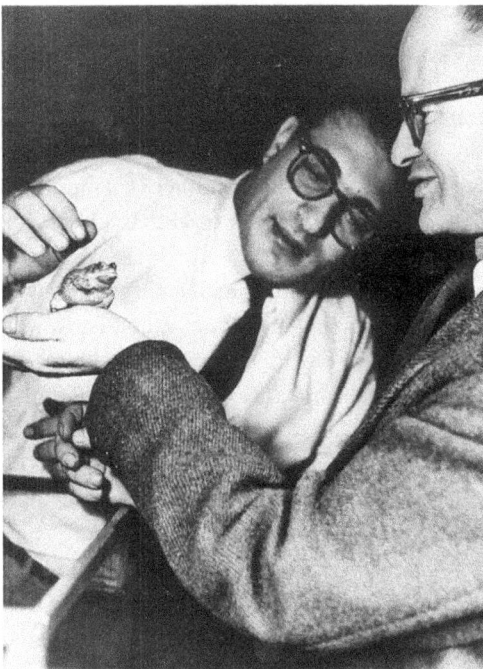

Figure B9.1. Jerome Lettvin (*left*) studying a frog with his brilliant friend and colleague Walter Pitts at MIT. (Author lapx86; Wikimedia Commons CC-BY-SA-3.0)

in the armed forces as a psychiatrist and then, following discharge, undertook specialist training in neurology. It was while working as a neurologist and night physician in a large mental health hospital south of Chicago that he began serious neurophysiological research on the mammalian spinal cord in his spare time. Despite claims of incompetence, he built the stimulators, amplifiers, and animal holders himself and soon began to attract the attention of other neuroscientists, among them Pat Wall. In 1950 Lettvin, together with Pat Wall and Warren McCulloch, were invited by Norbert Wiener to join the Research Laboratory of Electronics at MIT; there they were joined by a friend of Lettvin's from his teenage years—the extraordinary self-taught genius Walter Pitts. (It was Pitts who, with McCulloch, would write a hugely influential paper on the application of logical calculus to the understanding of neural networks, thereby creating a solid foundation for the development of artificial intelligence).* Of Lettvin's many research interests at MIT, two stand out. One is an exploration of field potentials in the mammalian spinal cord following peripheral nerve, or nerve root, stimulation—this work was a forerunner to more detailed studies on presynaptic inhibition by Pat Wall. The other major field of interest was vision in the frog, and it was this that led to the publication of a citation classic—"What the Frog's Eye Tells the Frog's Brain" (see main text).**

It was as a teacher, though, that Jerry Lettvin would be best remembered by many, and it during one of his lectures that he had come up with the extraordinary story of the (grand)mother cell.

* McCulloch W, Pitts W. A logical calculus of ideas immanent in nervous activity. *Bulletin of Mathematical Biophysics*. 1943; 5: 115–133.

**Lettvin JY, Maturana HR, McCulloch WS. What the frog's eye tells the frog's brain. *Proceedings of the IRE*. 1953; 47: 1940–1951.

tongues, mistaking them for bugs.[4] Clearly, in Lettvin's experiments, the image falling on the retina had undergone considerable analysis by the different types of cell in the retina before the ganglion cells discharged. In the mammalian brain, with its multiple visual areas in the cortex, there would be even greater opportunity for refinement of the features required to excite the cells.

This idea of convergence in a sensory pathway was not new, having been explicitly raised by William James as long ago as 1890:[5]

> There is, however, among the cells one central or pontifical one to which our consciousness is attached. But the events of all the other cells physically influence this arch-cell; and through producing their joint effects may be said to "combine."

Box 9.2 Lettvin's Fable

In the distant Ural mountains lives my second cousin, Akakhi Akakhievitch, a great if unknown neurosurgeon. Convinced that ideas are contained in specific cells, he had decided to find those concerned with a most primitive and ubiquitous substance—mother . . . And he located some 18,000 neurons that responded uniquely only to a mother, however displayed, whether animate or stuffed, seen from before or behind, upside down or on a diagonal or offered by caricature, photograph, or abstraction.

He had put the mass of data together and was preparing his paper, anticipating a Nobel prize, when into his office staggered Portnoy, world-renowned for his Complaint. On hearing Portnoy's story, he rubbed his hands with delight and led Portnoy to the operating table, assuring the mother-ridden schlep that shortly he would be rid of his problem. With great precision he ablated every one of the several thousand separate neurons and waited for Portnoy to recover. We must now conceive the interview in the recovery room.

"Portnoy?"
"Yeah."
"You remember your mother?"
"Huh?"
(Akakhi Akakhievitch can scarcely restrain himself. Dare he take Portnoy with him to Stockholm?)
"You remember your father?"
"Oh, sure."
"Who was your father married to?"
(Portnoy looks blank)
"You remember a red dress that walked around the house with slippers under it?"
"O Certainly."
"So who wore it?"
(Blank)
"You remember the blintzes you loved to eat every Thursday night?"
"They were wonderful."
"So who cooked them?"
(Blank)
"You remember being screamed at for dallying with shikses?"
"God, that was awful."
"So who did the screaming?"

(Blank)
And so it went . . . It made no difference—Portnoy had no mother. "Mother" he could conceive—it was generic. "My mother" he could not—it was specific . . .
Akakhievitch then . . . went back to . . . "grandmother cells."

Abridged from a letter Lettvin sent to Horace Barlow in 1995 and reproduced courtesy of Professor Barlow.

The great physiologist Sir Charles Sherrington had also considered the possible existence of "pontifical cells" in his authoritative *Man on his Nature*.[6] But having raised the possibility, Sherrington then rejected it, opting instead for a widely distributed population of neurons, each coding for a particular feature of the stimulus, and all their activities being integrated so as to give a conscious perception. And, indeed, there continues to be considerable support for a widely distributed scheme of this kind, one suggestion being that the cells involved would "bind" together by firing synchronous trains of impulses.[7,8]

Lettvin, however, responded to Barlow's suggestion with the fiction of Dr. Alexei Akakhievitch's curative surgery. It is important to note that cells relating to every aspect of the detested mother in the story were ablated, all several thousand of them. Thus there was no recollection of the mother wearing a red dress, or cooking blintzes, or screaming at her son. Every "mother" cell had been destroyed.

That such highly discriminating cells might actually exist in the mammalian brain was the discovery of Charles Gross and his colleagues in the Psychology Department at Harvard University. For a variety of reasons, they chose to explore the inferior temporal lobe;[9] until that time, little was known about the possible role of this region in vision. There was, however, one very important clue from neurology—patients with damage to this area, from a stroke for example, could become unable to recognize previously familiar faces (prosopagnosia). Nevertheless, Gross and his colleagues attempted to stimulate the inferior temporal cells not with faces but with bars, edges, and circles—much in the same way that Hubel and Wiesel performed their groundbreaking studies of the striate cortex (see Chapter 10). At the end of their paper, however, Gross and colleagues admitted that they might never have found the "best" stimulus for each neuron. There had been, for example, one cell that had responded best to a cut-out of a monkey hand—"The more the stimulus looked like a hand, the more strongly the unit responded to it." That the investigators seemed not to have appreciated the significance of their finding was all the more surprising because the senior author had recently reviewed a book that predicted the existence of just such cells—cells that would respond to whole objects rather than to their fragments (see later discussion).

The next advance came from Oxford where Edmund Rolls and his associates in the Department of Experimental Psychology had explored the temporal lobe with microelectrodes;[10] like Gross's team, they chose monkeys for their experiments but their target area was slightly higher up in the temporal lobe. Of the large number of cells studied, fully 10 percent responded best to human or monkey faces, regardless of their sizes or inclinations. The presence of face-selective cells in the inferior temporal lobe was soon confirmed by other laboratories.

Grandmother Cells ("Concept Cells") in the Human Hippocampus

What might well prove to be one of the last pages in the story of grandmother cells came in 2005 with a report in *Nature*[11] by a group based in Los Angeles that included Rodrigo Quiroga as well as Francis Crick's former collaborator, Christof Koch (Figure 9.2).[12] The authors had studied eight patients with

Figure 9.2. Kristof Koch. A close colleague of the late Francis Crick, Koch has written extensively on consciousness and is a major figure in the field. (2008 photo by Romanpoet, from Wikimedia Commons)

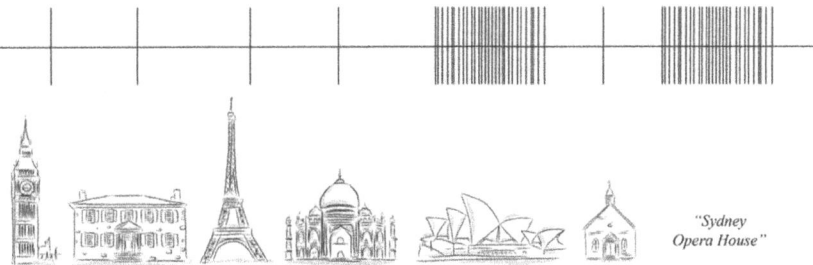

Figure 9.3. Diagram illustrating hippocampal neuron firing (*upper trace*) in response to the sight or name of a particular building, in this case the Sydney Opera House. (Based on observations of Quiroga et al.; see main text)

intractable epilepsy; it was hoped that electrophysiological recordings would aid in the identification of a lesion suitable for surgical resection. To this end, very fine wire electrodes were inserted through a common probe into different parts of the limbic system—the hippocampus, amygdala, entorhinal cortex, and parahippocampal gyrus. As in the previous studies, all of these structures were in the temporal lobe, but they differed in being on its medial surface (i.e., facing the midline). The subjects lay on their beds and faced a series of pictures, each of which was shown on a computer screen for one second; at the same time any impulse activity was recorded from the neuron under investigation. Remarkably, some of the cells proved to be highly selective in their responsiveness. One cell, for example, fired impulses whenever a picture of the film actress Jennifer Aniston was shown. Another cell, in a different patient, was stimulated by Halle Berry, responding not only to pictures of the actress but also to her printed name. Other less frequent cells appeared to be concerned with buildings rather than faces, one of them being fired both by pictures of the Sydney Opera House and by its spelled name (Figure 9.3). Because these various neurons responded to different aspects of the same object, be it person or building, the authors termed them "concept cells"—but they were, in every way, equivalent to the "(grand) mother cells" that had been the subject of Jerry Lettvin's fable.

Following this discovery, further information emerged about concept cells, largely through Rodrigo Quiroga (Figure 9.4) who has continued the California work following his move to Leicester and his collaboration with neurological and neurosurgical staff at King's College Hospital, London.[13] Thus the persons most often recognized by concept cells were found to be those important or familiar to the individual—not only family members but newly encountered medical or research staff. In the latter case the concept cells could only have been formed in the few days since admission of the patient to the hospital. Especially

Figure 9.4. Rodrigo Quiroga. A graduate in physics from the University of Buenos Aires, Quiroga became interested in the application of chaos theory to EEG and eventually in the properties of single neurons in the human brain. (Photo courtesy of Dr. Quiroga)

relevant for an understanding of conscious mechanisms was the observation that merely thinking about a person or image could increase the impulse firing rate of the corresponding concept cell, even when the person or image was no longer being seen. An indication of the hippocampal capacity for forming and storing memory was that concepts could be identified for only one to three percent of the neurons encountered.

Important and exciting as these new findings are, one of the interesting aspects of concept cells is that their existence had been predicted almost 40 years earlier, not only by Lettvin and Barlow but by a senior Polish neuroscientist as well—Jerzy Konorski.

Jerzy Konorski and *The Integrative Activity of the Brain*

Konorski's book appeared in 1967, published by the University of Chicago Press.[14] Although Konorski (Box 9.3) had spent some months at Yale working on the manuscript, the ideas expressed were his own and had been the outcome of years of observation and thought while serving as professor of neurophysiology

Box 9.3 Jerzy Konorski (1903–1973)

Jerzy Konorski, though one of the most penetrating and imaginative thinkers concerning higher brain mechanisms, remains largely unknown among Western neuroscientists. Born into a middle-class family in Lodz, Poland, in the early years of the 20th century, Konorski followed his schooling by studying medicine at Warsaw University. Hoping that he might learn how the brain functioned and having discovered the publications on conditioned reflexes by Pavlov in Leningrad, St. Petersburg, Konorski and his fellow student Stefan Miller decided to carry out some experiments of their own. Eventually, in a small upstairs room with a dog bought in the local market and with toilet paper as the cheapest material for their drum kymograph recordings, they started their studies. They found they could produce conditioned reflexes with a motor response (paw raising), as opposed to the salivary secretion employed in all the Pavlovian studies. After graduation, Konorski and Miller were invited to join Pavlov's group in Leningrad, Konorski staying for two years. Pavlov, however, maintained that their type of conditioned reflex was no different from his own classical model. Back in Warsaw, Konorski was appointed to the Nencki Institute and met and married Liliana Lubinska,

Figure B9.3. Jerzy Konorski. A Polish neurologist, psychologist, and neuroscientist—and a man with highly original ideas regarding brain function. (Author unknown; Wikimedia Commons)

who had trained in neurophysiology in Louis Lapique's laboratory in Paris. Further studies in Warsaw were interrupted by the start of World War II. Prevented from fleeing to the United Kingdom, Konorski—through the influence of friends in Leningrad—became head of the physiology department at the Primate Biological Station at Sukhumi on the Black Sea, before being evacuated to the Georgian capital, Tiblisi.

Following the end of the war in 1945, Jerzy Konorski returned to a devastated Poland, a country that had lost many of its finest scientists, including his friend and former colleague, Stefan Miller. With great determination Konorski and others recreated the Nencki Institute in Lodz, before moving the institute to a new building in Warsaw. In 1948 Konorski's first book, *Conditioned Reflexes and Neuron Organization*, was published.* After further work on reflexes, Konorski switched to studying the effect of cortical lesions in animals and neurological patients, with assistance from a neurosurgeon, Lucjan Stepien. Following a three-month invited visit to the United States in 1957, Konorski's work became more widely known, and he was encouraged to write his second book, *Integrative Activity of the Brain*. Despite having an American publisher,** the book, which appeared in 1967, failed to achieve the impact that Konorski had hoped for. Konorski himself attributed the poor response to the dense writing required to express all his ideas. However, his concept of "gnostic cells" has been confirmed,*** as has his speculation that these neurons would be found in the temporal lobe.

Source. konorski.nencki.gov. pl.
* Cambridge University Press.
** University of Chicago Press.
*** The present awareness of Konorski's work among Western neuroscientists is largely due to the influence of Charles Gross, former head of psychology at Harvard and then at Princeton. See: Gross CG. Genealogy of the "grandmother cell." *Neuroscientist.* 2002; 8: 512–518.

at the Nencki Institute in Warsaw. There was little reference to electrophysiological recordings in the book, other than to the relatively recent ones of Hubel and Wiesel; instead Konorski drew on his extensive knowledge of reflex behavior in humans and animals and on the clinical effects of damage to different areas of the brain. The book was written in impeccable English, was strikingly original in showing how the brain might work, was favorably reviewed in *Science*—and yet was allowed to gather dust with hardly anyone in the West noticing it![15]

One of Konorski's conclusions, when considering sensory perception, was that there must be grandmother cells—except that he called them "gnostic units" instead. He envisaged the various sensory pathways as "analyzers," with each analyzer comprising a series of levels, the lowest of which were the receptors (Figure 9.5). Those neurons passing information on to the next higher level were

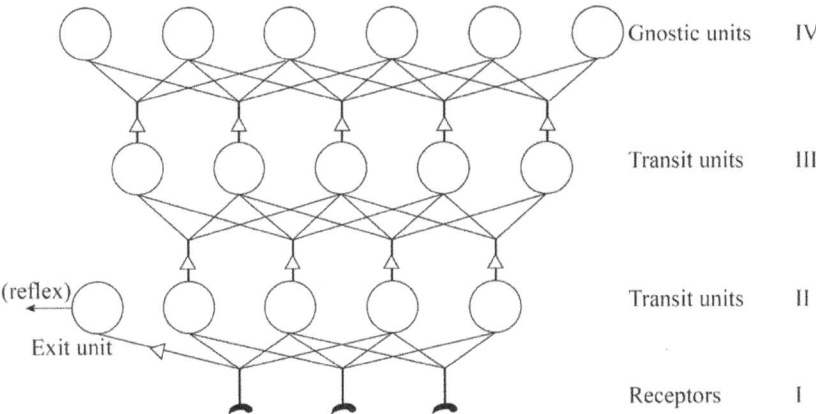

Figure 9.5. Konorski's hierarchical scheme for the functioning of a sensory "analyzer."

"transit units," while those that sent axons locally, perhaps as part of a reflex, were "exit units." The higher the level of the sensory pathway, the more precise the stimuli required to excite the transit units. At the top of this hierarchical scheme, in the cerebral cortex, were the "gnostic units"; these were the neurons that, when excited in the visual system, were responsible for the final image—a face perhaps, or an animal, or some other object in a different category. Importantly, the elements in the visual field that were adequate to excite the transit units, such as Hubel and Wiesel's bars, slits, and edges, were not seen as such in the final picture.

In Konorski's scheme the perceptions resulting from activation of the gnostic neurons were "unitary" and did not depend on associations. For example, a face could be recognized even though the name of its owner did not immediately come to mind. The gnostic units had a certain robustness in that, for example, a face could be identified despite being altered in some small way—by the wearing of spectacles, perhaps. Similarly, a word could be read correctly, even though there might be a letter missing. It was that particular word and that particular face that were the unitary events. From clinical observations Konorski deduced that the gnostic units were situated not in the primary cortical receiving areas (where elementary features of the stimulus could be mapped out by the neurophysiologist) but in the association areas—including the temporal lobe. If large enough, a localized lesion in such an area (caused by an injury, tumor, or stroke) could result in an inability to recognize not just individual stimuli but an entire category—for example, faces (prosopagnosia), letters of the alphabet (alexic agnosia), or objects placed in the hand (astereognosis). It was presumably a lesion of this kind that had led to one of Oliver Sacks' patients being unable to distinguish his wife from a hat.[16]

Konorski further suggested that, for a gnostic unit, and hence an elemental memory, to be formed in the brain, not only must there be impulse activity proceeding through the different levels of the transit ("cognitive") system to the association cortex but also an "emotive drive"—that is, an input from a system comprising the hypothalamus, reticular formation, nonspecific thalamic nuclei, and the limbic system. It was this combination that made potential synaptic connections into functioning ones. A young brain would have a greater number of possible gnostic units than an old one. Although a gnostic unit did not depend on associations, as noted previously, such a linkage between different categories of units would normally tend to occur. For example, the face of a person would be associated with his or her name, with that person's voice, and perhaps with the clothes that that person usually wore or the house that he or she lived in.

This explanation is merely part of Konorski's conception of the brain processes that underlie perception. The important point is that he deduced the presence of neurons, each of which represented a complete entity, be it a person or an animal, tree, car, house, written word, or something else. This thinking was the very opposite of the widely held supposition that a perception resulted from the combined activities of a relatively large number of neurons, each of which coded for a different part of the object viewed (or heard, touched, smelled). Where Konorski stopped short was in not suggesting that the synaptic linkages between neurons representing different aspects of the same person or object (visual image and name, for example) could be so strong as to be combined in yet higher order gnostic units—the "concept" or "grandmother" cells.

"Place Cells" in the Rat Hippocampus

Before leaving Konorski's ideas, it is instructive to speculate about the possibility of gnostic units in brains simpler than those of humans or monkeys. What is the situation in a rat brain, for example—for this is an animal that, unlike humans, has very poor vision and is unlikely to be interested in faces and in the sorts of objects encountered by humans? Nevertheless, despite the small size of the rat brain, there is a hippocampus in each hemisphere. So what does it do?

The answer, possibly a partial one, came from experiments conducted by John O'Keefe (Box 9.4). O'Keefe, an American, joined Patrick Wall at the latter's Cerebral Function Research unit at University College London in the early 1970s. Wall (Box 9.5) became prominent for his pioneering work on presynaptic

Box 9.4 John O'Keefe (1939–)

Born in Harlem, New York, John O'Keefe is the son of working-class Irish immigrants who suffered during the Depression. John, growing up in the South Bronx, proved to be an undistinguished student at his Jesuit school and, following graduation, was obliged to work first as a clerk and then as a bookkeeper, all the while attending evening classes in aeronautical engineering at New York University. O'Keefe was then able to obtain a position with the Grumman Aeronautical Company on Long Island but subsequently decided to become a full-time student at City College of New York. While taking a course in physiological psychology he had an opportunity to take part in experiments on pigeon behavior and developed an enthusiasm for investigating the brain. He next enrolled at McGill University as a graduate student in the prestigious psychology department headed by Donald Hebb.*

Figure B9.4. John O'Keefe. Clearly elated to have won the Nobel Prize! (Author, Per Henning/NTNU; Wikimedia Commons CC BY 2.0)

His PhD advisor was Ronald Melzack** and his research project involved recording from single neurons in the rat amygdala.

In 1967 O'Keefe moved to the United Kingdom as a postdoctoral fellow in the Cerebral Function unit of the Anatomy Department at University College London. There, under the direction of Patrick Wall, he began to study neurons in the somatosensory systems of freely moving rats, initially in the dorsal column nuclei and then in the thalamus. By chance, a misplaced electrode in the hippocampus yielded a cell with very interesting properties, its firing was linked to movements of the animal and to a local theta rhythm. There and then O'Keefe decided to switch his attention to the hippocampus, doing so with Wall's full support. The new project soon paid dividends with the discovery of some cells that were responsive to the speed of movement of the rats and others to the animal's position in two-dimensional space (the "place" cells). The combined activities of the two types of cell, and of others discovered later, evidently enabled the rats to construct cognitive maps of their surroundings and of their own locations. This work was developed fully and imaginatively with a number of visiting scientists, including May-Britt and Edvard Moser, who joined O'Keefe in sharing the 2014 Nobel Prize in Physiology or Medicine. John O'Keefe is now the founding director of the Sainsbury Wellcome Centre for Neural Circuits and Behaviour at University College London

Source. https:www.nobelprize.org/nobel_prizes/medicine/laureates/2014/okeefe-bio.html

* Donald Hebb, a Canadian, was one of the most prominent experimental and theoretical psychologists of his day and one very interested in the way that memories were laid down. In his influential book *The Organization of Behavior: A Neuropyschological Theory* he postulated that repeated excitation strengthened synaptic connections. See Chapter 5 for more information on Hebb.

** Melzack achieved prominence for his work on phantom limb sensations and for his studies with Patrick Wall on pain mechanisms; the two postulated a neural "gate" in the dorsal horn of the spinal cord, which controlled the flow of impulses in ascending pain pathways.

inhibition and pain. At the time of O'Keefe's move, single unit studies in rats were not uncommon, one of the first exponents being George Dawson,[17] also at University College London. Unlike cats and monkeys, rats were cheap and plentiful, and the ethical issues associated with their exploitation were smaller; moreover, countless experiments in psychology laboratories had shown rats to be good learners. O'Keefe devised a simple but effective way of recording from single cells in the brains of unrestrained rats using adjustable implanted platinum wire electrodes. Following up on a chance recording of unusual interest, he and his graduate student Jonathan Dostrovsky discovered that among the hippocampal

Box 9.5. Patrick Wall (1925–2001)

Although strongly eschewing any form of religion, Patrick Wall nevertheless had elements of a biblical prophet, an impression heightened by the luxuriant beard and the gravity and force of his statements. Like a prophet he spoke out against false gods (especially authoritarian heads of departments) and, again like a prophet, he could emerge with manna after years of toil in the wilderness (the substantia gelatinosa in the spinal cord). Indeed, it seemed that, from wherever he turned his attention, new ideas—radical ones at that—were sure to follow.

Born in Nottingham, England, Wall went to Oxford with the intention of ultimately obtaining a medical degree. Like a number of other potential physicians or surgeons, however, he became interested in basic science, which in his case was neuroanatomy. Under the guidance of Paul Glees, Wall

Figure B9.5. Patrick Wall attending a conference in Kingston, Ontario, in 1995. As ever, with a cigarette. (Author's photograph)

studied fiber pathways in the nervous system by staining the degenerating terminal branches of sectioned axons. After qualifying in medicine, Wall worked briefly at Yale in John Fulton's department, before transferring to the University of Chicago. The attraction was the presence of the youthful Jerry Lettvin, a full-time physician who had somehow succeeded in building and operating a neurophysiology laboratory in the psychiatric hospital in which he worked. When Lettvin and Warren McCulloch moved to MIT to set up a neurocomputational research unit, Wall went with them. There, in Boston, he began a source-sink analysis of the sensory potentials evoked in the dorsal horn of the spinal cord, a novel line of work that led to the discovery that impulse activity could be controlled presynaptically. There was also recognition of the key role of the substantia gelatinosa in controlling sensory inflow, a concept that spawned the hugely influential and widely accepted gate theory of pain (published with Ronald Melzack), and the subsequent employment of repetitive electric stimulation for pain relief.

In 1967 Wall returned to the United Kingdom to join J. Z. Young, the renowned zoologist and head of anatomy at University College London. There Wall established a cerebral functions research unit, a feature of which was the recording of impulse activity in freely moving rats by means of implanted electrodes—a technology also used by his junior colleague and later Nobelist John O'Keefe. Among other achievements, Wall identified "silent" synapses, spontaneous impulse activity in regenerating nerve fibers, and rapid-onset neuroplasticity after lesions.

An ardent socialist and one-time communist, Wall was nevertheless opposed to centralized authority, a philosophy that carried over into his own research. Long after his official retirement from University College London, Wall continued to experiment and teach, his lectures often applauded by those most cynical listeners, medical students. Though a life-long chain smoker, it was cancer of the prostate to which he would eventually succumb.

cells there were some that fired during a particular activity—sniffing, for example—but there were others that only became active when the rat was in a certain position inside a maze.[18] Thus, whenever the animal came back to that location, the same cell(s) would begin to discharge and would continue to do so until the rat moved off. Other cells were found later that only responded to the nearness of walls and edges, and, from experiments using a water maze, it became clear that vision was important in providing the necessary information. From the combined activities of these various types of cell, a map of the rat's environment was evidently being constructed in the hippocampus.[19] Was it possible, then, that the place cells, the activity of which provided memories of importance in the life of the rat, were the equivalent of grandmother cells (gnostic units) in humans?

Next we inquire how a sensory system, such as the visual one, might operate.

Notes and References

1. This was first done in the author's laboratory. See: McComas AJ. Fawcett PRW, Campbell MJ, Sica REP. Electrophysiological estimation of the number of motor units within a human muscle. *Journal of Neurology, Neurosurgery and Psychiatry*. 1971; 34: 121–131.
2. Barlow HB. Summation and inhibition in the frog's retina. *Journal of Physiology*. 1953; 119: 69–88.
3. Barlow HB. Single units and sensation: a neuron doctrine for perceptual psychology. *Perception*. 1972; 1: 371–394.
4. Lettvin JV, Maturana HR, McCulloch WS, Pitts WH. What the frog's eye tells the frog's brain. *Proceedings of the IRE*. 1959; 47: 1940–1951.
5. James W. *The principles of psychology*. New York, NY: Dover, 1890.
6. Sherrington CS. *Man on his nature*. Cambridge, England: Cambridge University Press, 1940.
7. Singer W. Synchronization of cortical activity and its putative role in information processing and learning. *Annual Review of Physiology*. 1993; 55: 334–350.
8. Crick F, Koch C. A framework for consciousness. *Nature Neuroscience*. 2003; 6: 119–126
9. Gross CG, Bender DB, Rocha-Miranda CE. Visual receptive fields of neurons in inferotemporal cortex of the monkey. *Science*. 1969; 166: 1303–1306.
10. Perrett DI, Rolls ET, Caan W. Visual neurons responsive to faces in the monkey temporal cortex. *Experimental Brain Research*. 1982; 47: 329–342.
11. Quiroga RQ, Reddy L, Kreiman G, Koch C, Fried I. Invariant visual representation by single neurons in the human brain. *Nature*. 2005; 435: 1102–1107.
12. Christof Koch is now president and chief scientific officer of the Allen Institute in Seattle. As a scientist, his specialty is in neural computation. Included among his many publications is an autobiography: *Consciousness. Confessions of a Romantic Reductionist* (Cambridge, MA: MIT Press, 2012).
13. Rey HG, Ison MJ, Pedreira C, et al. Single-cell recordings in the human medial temporal lobe. *Journal of Anatomy*. 2015; 227: 394–408.
14. Konorski J. *Integrative activity of the brain: an interdisciplinary approach*. Chicago, IL: University of Chicago Press, 1967.
15. Though copies may still be found in university libraries, few are available for sale through the Internet. It is only now, largely through the efforts of Charles Gross, that the importance of Konorski's ideas and his remarkable foresight are finally being recognized.

16. Sacks O. *The man who mistook his wife for a hat and other clinical tales*. New York, NY: Simon & Schuster, 1985.
17. For more on Dawson, see next chapter.
18. O'Keefe J, Dostrovsky J. The hippocampus as a spatial map: preliminary evidence from unit activity in the freely moving rat. *Brain Research*. 1971; 34: 171–175.
19. O'Keefe J. *Spatial cells in the hippocampal formation*. Nobel Lecture, December 7, 2014. nobelprize.org.

10
Making Sense of the Senses

We live by our senses. Everything that we know of the world comes to us through them, so much so that we have used our senses to ascribe colors and shapes, sounds, tastes and smells to the people and objects around us. Yet the reality is that, of themselves, the people and the objects have none of these qualities. It is the visual system that is, for example, able to translate the photons of a particular wavelength into the redness of a rose or a girl's dress. Similarly, it is the frequencies of the pressure waves emanating from the sounding board of the piano and striking the eardrum that are interpreted by the auditory system as the repeated triplets of Beethoven's "Moonlight Sonata." Thus, improbable as it may at first appear, neither the dress has color nor the piano sound; rather it is our brains that give them these attributes. How these transformations come about—how electrical impulses in the brain give rise to redness or to the imperiousness of a Beethoven chord—is still a mystery and is likely to remain so. Indeed, this is the "hard" problem of consciousness that the philosopher David Chalmers, among others, has drawn attention to. On the other hand, even if we cannot comprehend the final step, there is now much information as to how the brain processes information from the eye and ear, and from the other sensory receptors, on its way to consciousness.

Of the various sensory systems, the one about which most is known is the visual one. Indeed, it was the abundance of knowledge concerning vision that led Francis Crick to choose that sensory modality as his portal into the study of conscious mechanisms. Though vision is also the most complex sensory system, reflected in its large cortical territory, it is the one that we consider next. Some of the principles at play in vision may, however, be a feature of other sensory systems, and of somatic sensation in particular.

The Eye

The complexity of the visual system begins with the eye. It is the intricate construction of the human eye that has sometimes been used as an argument for the refutation of evolution and thus for the presence of Intelligent Design in the universe. And yet the mammalian eye did evolve, though it required a multitude of steps to proceed from a solitary photoreceptor in a single-celled organism to do so.

However, it was not until the brilliant anatomical studies of the retina by Santiago Ramon y Cajal and his contemporaries that it became possible to appreciate the full complexity—and beauty—of the eye. A talented self-taught artist, Cajal conveyed his findings with the Zeiss light microscope into a series of beautiful pen and ink sketches of nerve cells and fibers. Prominent among the latter was his depiction of the arrangement of the nervous elements in the retina (Figure 10.1).

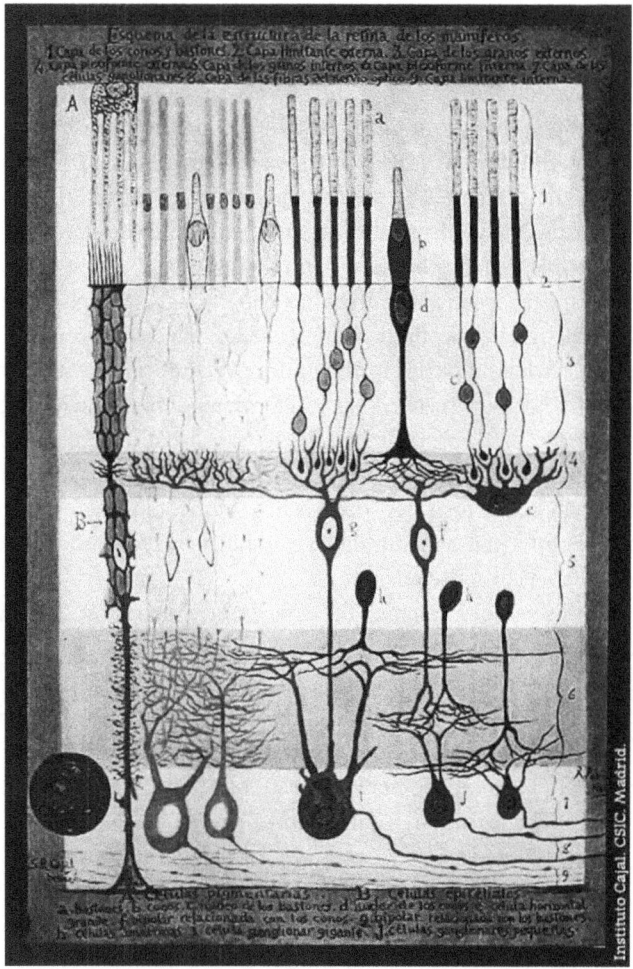

Figure 10.1. Structure of the mammalian retina. Ink sketch by Ramon y Cajal, c. 1900. The original is even more impressive, as different colored inks were used. The various cell types identified by Cajal include the rods (*a*) and the less numerous cones (*b*) in the photoreceptor layer (*1*), bipolar cells (*f, g*), horizontal cell (*c*), amacrine cell (*h*) and ganglion cells (*i, j*). (Wikimedia Commons)

To understand how the different types of cell in the retina work together and to follow the resulting impulse activity further up the visual pathway, there is no better account than that given in David Hubel's monograph, *Eye, Brain and Vision*.[1] For present purposes it is enough to say that daylight vision is mediated by cones that respond directly to the photons; via the bipolar cells, the electrical output from the cones ultimately reaches the ganglion cells, the fibers of which form the optic nerve. It is the ganglion cells, then, that provide the brain with the information it needs to construct the image of what the eye has seen. But in what form is this information from the ganglion cells? To answer this question, it was necessary to make recordings from individual ganglion cells in the retina with microelectrodes.

Stephen Kuffler's Study of Ganglion Cells

Of the various investigations of the ganglion cells, none was more important than that of Stephen Kuffler (Box 10.1). Hungarian by birth, Kuffler fled Europe prior to the outbreak of World War II and ended up in the neurophysiology laboratory of John Eccles in Sydney, Australia. There, also in the company of future Nobel Laureate Bernard Katz, Kuffler taught himself the delicate dissecting techniques that would enable him to develop novel preparations for answering a wide range of neuroscientific questions later in his career.[2] In 1950, however, Kuffler was found in the basement of the Wilmer Institute of Ophthalmology at the Johns Hopkins Hospital in Baltimore, Maryland. Kuffler had proceeded to do the logical thing for a physiologist in his new situation—study how the visual system functioned.

Using the cat for his experiments, Kuffler shone very small spots of light on to different regions of the retina and found that there were two types of ganglion cell.[3] One type ("center-on") fired impulses when the spot was on a certain small part of the retina and ceased firing when the spot moved to the surrounding area (Figure 10.2). The other type of cell ("center-off") did just the opposite, being inhibited in a central area and excited in the surround. Both types of cell, however, had one thing in common—their excitatory and inhibitory areas on the retina ("receptive fields") were concentric.

There was an additional characteristic that could be demonstrated in a proportion of the ganglion cells, and this was a consequence of the different color sensitivities of the cones in the retina. This additional property was "color opponency." For example, a ganglion neuron might respond to red light falling on the center of its field in the retina and be inhibited by green light striking the surround and vice versa. Or, in other ganglion cells, there could be opponency

Box 10.1 Stephen Kuffler (1913–1980)

Like Ramon y Cajal before him, Stephen Kuffler would have seemed an unlikely person to attain the highest honors in science. Kuffler was born at the height of the Austro-Hungarian Empire in the village of Táp in Hungary, where his father was an affluent landowner. Following the start of the communist dictatorship in post–World War I Hungary, Kuffler's family fled to Austria where, after education in a Jesuit boarding school, Kuffler chose medicine as a career. Later, Kuffler moved back to Hungary, then visited England, and finally, alarmed by the rise in Nazism, sailed to Sydney, Australia—this without any job prospects. By then Kuffler had changed his first name from "Wilhelm" to "Stephen," and it was as Dr. Stephen Kuffler that he was taken on as an unpaid lecturer and assistant to John Eccles in the Department of Pathology at the University of Sydney. The first sign of Kuffler's remarkable abilities came

Figure B10.1. Stephen Kuffler at Harvard. Kuffler's perpetual good humor made him well liked by his colleagues and the many doctoral and postdoctoral students that trained under him.
(Photograph courtesy of Dr. U. J. McMahan)

when, after considerable practice, he succeeded in preparing a single motor axon-muscle fiber preparation for studies on the neuromuscular junction.

At the end of the war, Kuffler accepted an invitation to work in Ralph Gerard's physiology department at the University of Chicago and there studied the γ-motor innervation of the muscle spindle, again using his self-taught dissecting skills. Kuffler next moved to Baltimore where he was given a laboratory in the Wilmer Institute of Ophthalmology at Johns Hopkins Hospital. While investigating the retina, Kuffler showed that the ganglion cells responded best to small spots of light, especially those with dark surrounds. It was during this phase of his career that Kuffler began to attract young neuroscientists, including the future Nobel Laureates David Hubel and Torsten Wiesel. In summers spent at Wood's Hole, Massachusetts, Kuffler exploited his dissection of the crayfish stretch receptor to describe the generator potential, presynaptic inhibition, and the role of GABA as an inhibitory neurotransmitter. Indeed, Kuffler had a knack of finding the best preparation to suit his purpose, whether it was frog, crayfish, leech (for its glial cells), or something else.

In 1959 Kuffler took his neuroscience team to the Department of Pharmacology at Harvard and, his success continuing, was given his own department in 1966. The title—Department of Neurobiology—reflected the fact that the department embraced not only neurophysiology but also neurochemistry, electron microscopy, and, later, molecular biology. Still at the height of his powers, Kuffler fell victim to a massive heart attack following his daily swim at Wood's Hole. Had he not done so, he might well have shared the following year's Nobel Prize with his former protégés, David Hubel and Torsten Wiesel.

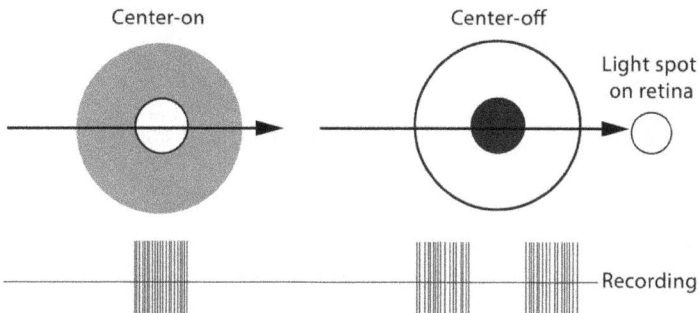

Figure 10.2. Center-on (*left*) and center-off cells. As a spot of light (arrowed) moves across the respective receptive fields on the retina, the impulse discharge increases or decreases, as shown below.

between blue and yellow. For many of the ganglion cells, however, there was no opponency other than that for light and dark.[4]

One might have thought, intuitively, that the information conveyed by impulses in the fibers of the retinal ganglion cells would undergo some kind of processing when they reached the next stage in the visual pathway, the lateral geniculate nucleus (lying immediately posterior to the thalamus). This proved not to be the case, however—the receptive fields on the retina were still concentric, with some of them being "center-on" and the remainder "center-off." Just as with the ganglion cells in the retina, a substantial fraction of the geniculate cells exhibited color opponency.

Hubel and Wiesel's Exploration of the Visual Cortex

Compared with the rather unexciting results in the lateral geniculate, the investigation of the next stage in the visual pathway, the primary visual receiving area and its neighbor in the occipital cortex (areas V1 and V2, respectively), proved far more rewarding. Indeed, the findings were so unexpected that they, and their implications as to how the visual brain worked, resulted in Nobel prizes for their discoverers, Torsten Wiesel (1924–) and David Hubel (1926–2013; Figure 10.3). Wiesel, a Swede, graduated in medicine at the Karolinska Institute

Figure 10.3. David Hubel (*left*) and Torsten Wiesel examining the properties of a complex cell. (Reproduced courtesy of the late Dr. David Hubel)

and then, after teaching physiology for a year, joined Stephen Kuffler in Baltimore to gain research experience. In 1958 David Hubel appeared. Born in Canada of American parents, Hubel obtained his medical degree at McGill University (where, by sharing his lecture notes, he helped his much older fellow student, Herbert Jasper, obtain his degree). In 1954 Hubel moved to Johns Hopkins in Baltimore for a residency in neurology. Then, drafted into the army, he began research on the visual cortex, developing the tungsten microelectrode, together with a hydraulic microdrive, for recording from single neurons. Hardly had Hubel settled down to continue his work, now in collaboration with Wiesel, then Kuffler accepted an invitation to form a neuroscience research unit at Harvard University, taking Hubel and Wiesel to Boston with him.

The move to Boston took place in 1959. Since the end of World War II Harvard had lost two of its most accomplished neuroscientists. Hallowell Davis (1896–1992), perhaps the world's leading authority on the physiology of hearing, had taken his knowledge and expertise to St. Louis. In addition to his extensive studies of the auditory system, Davis had been the first to record the EEG alpha rhythm in the United States while his wife, Pauline, had been the first to record auditory-evoked potentials in human subjects. The other loss was Alexander Forbes (1882–1965), whose numerous prewar publications had dealt with nerve conduction, excitation and inhibition, spinal reflexes, EEG, and late cortical evoked responses. Now an emeritus professor, Forbes had moved to the biology laboratories at Harvard and was studying color vision in turtles. However, it was not to the Physiology Department that Kuffler came but to the Department of Pharmacology. Seven years later, in 1966, he was given his own department—the world's first Department of Neurobiology. Following Harvard's example, other universities would create departments of neuroscience or neurobiology, sometimes out of anatomy and physiology.

As Kuffler had done with the lateral geniculate nucleus, Hubel and Wiesel used small spots of light to begin their exploration of the cat primary visual cortex (V1). Rather surprisingly, the stimuli, which had been so effective in the geniculate, evoked meager responses in the cortex. And then, not for the first time in neurophysiological research, chance intervened. As the investigators entered a new slide into the projector used for stimulating the retina, the cortical cell they were recording from gave a brisk discharge. Intrigued by this unexpected result, Hubel and Wiesel discovered that the cell had been stimulated by the moving edge of the slide as it slipped into place. Pursuing this observation, Hubel and Wiesel discovered that linear stimuli—lines, slits of light, edges—were much more effective means of exciting visual neurons than spots of light.[5,6] There were other fundamental observations. First, the linear stimulus had to have a particular orientation in the visual field (and retina) to be maximally effective. Second, the same optimal orientation applied to other neurons in a column perpendicular to the surface of the cortex. This was not altogether surprising, however, for the same principle of columnar organization had been observed previously, by Vernon Mountcastle in his exploration of the somatosensory cortex.[7] Earlier still, it had been proposed by Rafael Lorente de Nó on the basis of his anatomical studies.[8] The third finding of Hubel and Wiesel was that there were inhibitory surrounds to the excitatory linear stimuli.

Hubel and Wiesel termed the visual neurons that exhibited this behavior "simple cells" and thought it probable that each one received inputs from a small number of geniculate cells whose small concentric visual fields formed a line (Figure 10.4). "Complex cells" tended to be more numerous, and each received projections from several simple cells. Instead of responding to stationary lines in a certain visual territory, they were excited by the movement of lines (or bars or edges) across the visual field—provided the lines had a critical orientation. The incidence of complex cells was higher in area V2 and higher still in V3, both areas part of Brodmann area 18. Hubel and Wiesel also discovered that the projections from the two eyes preferentially innervated alternating bands of neuron columns in the primary visual cortex ("ocular dominance columns"). For

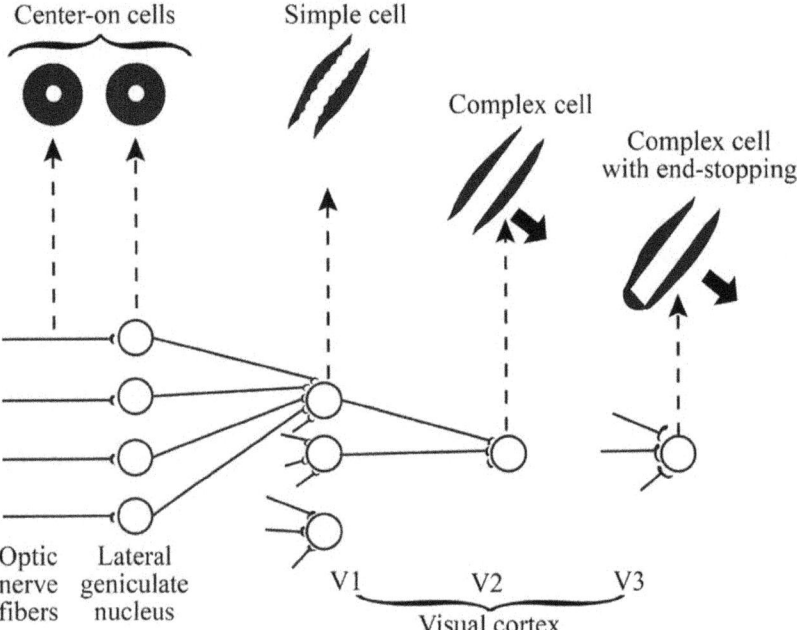

Figure 10.4. Hierarchy of visual neurons. On reaching the primary visual cortical area (V1) the axons of four lateral geniculate center-on cells (*left*) converge on to a simple cell that, in turn, converges with two other simple cells on to a complex cell.

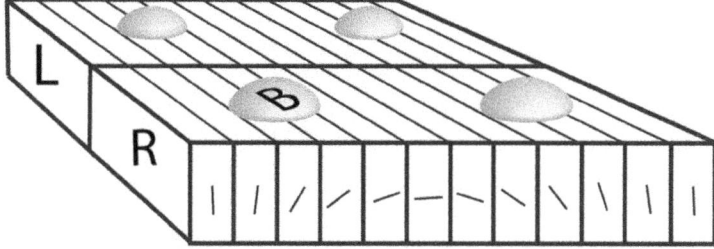

Figure 10.5. The ice-cube model of columnar organization in V1 proposed by Hubel and Wiesel. Each rectangular block represents columns of cells responsive to a particular orientation of a linear stimulus in the visual field (orientation shown by line at front). Intermingled with the columns are "blobs" (*B*) of color-sensitive neurons with concentric receptive fields. Alternate rows of columns receive information predominately from left (*L*) or right (*R*) eyes, respectively.

example, if one column of neurons responded most strongly to stimuli from the left eye, a column in a neighboring band would be excited best by inputs from the right eye serving the same part of the visual field. Hubel and Wiesel were able to summarize the organization of the neurons in the primary visual cortex in their so-called ice-cube model (Figure 10.5).[9]

Parallel Processing in the Visual Cortex

Hubel and Wiesel did much more, including studies on the development of ocular dominance in the newborn animal, but it was the transition of effective stimuli from circles to lines in the primary visual area that attracted the most attention. The results suggested that the primary visual cortex (V1) began processing information from the eyes by recognizing linear features in the visual field and that this activity was then refined in V2 and V3 as part of a functional hierarchy. By implication, further processing in more anteriorly situated cortex, this time in the temporal lobe, would eventually yield the perceived image. However, other evidence, much of it obtained by Semir Zeki in University College London, suggested that the analysis of the visual field might be a parallel, rather than a serial, process.[10] Thus the perception of color was considered to be carried out separately in area V4, partly because of the responses of neurons to light of different wavelengths and partly because patients with lesions (usually strokes) confined to this area could lose all sense of color in the affected part of the visual field. Another cortical area, this time in the middle temporal lobe (V5), was considered to deal with movement in the visual field. Again, the evidence was partly clinical since, after

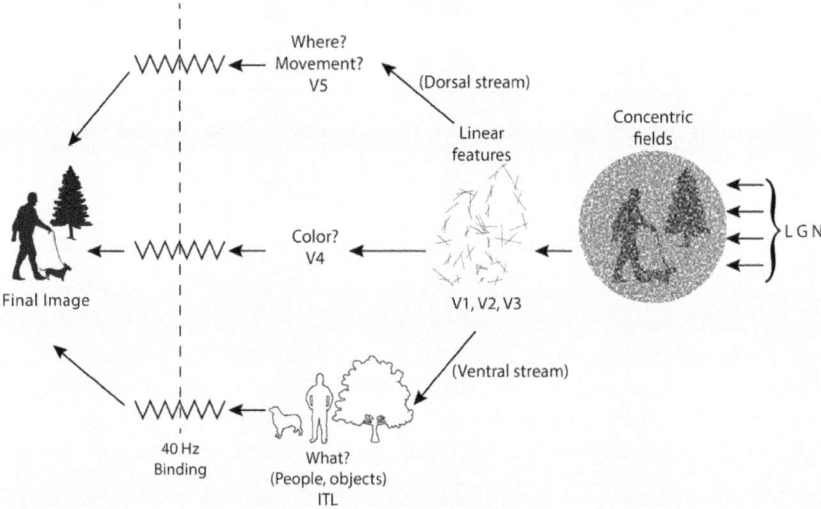

Figure 10.6. Visual pathways involved in parallel processing scheme.

damage in this area, movements were no longer seen to flow smoothly but in jerks (as in the pouring of tea from a teapot, for example).[11]

It seemed, then, that different parts of the brain dealt simultaneously with different aspects of the visual image. Later studies suggested that there were two main pathways. A dorsal stream, running from V1 in the occipital cortex forward into V5 in the middle temporal cortex, was considered to deal with movements and "where" in the visual field (Figure 10.6). A ventral stream, running from V1 to V2 to V3 and then into the inferior temporal cortex, was thought to be concerned with object recognition, that is, the "what." Color analysis was the function of V4.

Accepting such a division of tasks raised the question as to how the different parts of the analysis would be combined. For example, how could the activities of the neurons detecting linear features become integrated with those responsible for colors and motion? One possible way of dealing with this problem was put forward by Christof von der Malsberg in 1981 and then taken up by Wolf Singer[12] and, later still, by Francis Crick and Christof Koch.[13] The suggested solution was that the activities of the various processing areas were "bound" together by means of synchronous impulse firing in the 30 to 60 Hz (gamma) range. However, while it was certainly true that neighboring neurons tended to discharge together, especially in the presence of a slow wave, it was more difficult to demonstrate the phenomenon for cells at a distance from each other. Further, there was good evidence from masking experiments, considered later, that rather than sensory consciousness being a smooth, continuous process, it was actually broken up into "chunks" of time. Since each chunk lasted 100 ms or less, there would have been insufficient time for more than a very few synchronous discharges between distant cortical areas. There was also the issue of "saccades," the continual small flickering movements of the eye from one part of the visual field to another (see later discussion); there would hardly be time for an impulse frequency code to become established before new information from the retina arrived.

An Alternative Processing Scheme

While the Hubel and Wiesel scheme of linear analysis, followed by distributed processing in the visual areas, was widely accepted, there were a number of concerns, quite apart from the binding problem. One of the concerns[14] was that, while simple and complex cells were obviously suited for outlining structures such as doors, windows, walls, and posts, they would be very inefficient in analyzing the shapes of rounded objects, especially if they were small (coins, buttons, pills, lowercase lettering, etc.). Further, if the simple and complex cells were so important in image analysis, their incidence among other types of cells would be expected to be highest in the relatively large area of visual cortex dealing with the central 1% to 2% of the visual field. This, the most important part of the visual field, is subserved by the fovea—that small region of retina with the highest concentration of cones (enabling the greatest acuity). This was found not to be the case, however; the incidence of orientation-selective cells was observed to be higher outside than inside the foveal region of V1 cortex.[15] Another argument against the involvement of the orientation-selective cells in feature analysis was that their receptive fields would be shortest in the foveal representation, making them less effective as linear detectors.

There were other arguments against orientation processing in the retina, and one of the most persuasive came from clinical neurology and concerned individuals suffering from migraine.[16] Many such patients commonly experience a visual aura before the onset of their headache. The most common aura consists of a flickering ring of zigzags (the "fortification spectra") that slowly expands across the visual field, leaving temporary dimness of vision in its wake (Figure 10.7). If each zig or zag was due to the spontaneous firing of a column of simple or complex cells in the cortex, as was highly likely,[17] then the consistent angulation of their receptive fields would make them of very limited use as linear detectors (Figure 10.8). Also, and perhaps more tellingly, it is sometimes possible for a migraineur to see objects in the visual field without any impairment during an aura, despite the fact that the simple and complex cells are presumably engaged in producing the zigzags. So are there any other possible schemes for visual processing?

The answer is yes. Supposing that the simple and complex cells do not contribute directly to object recognition, then the only other candidates for this task are the

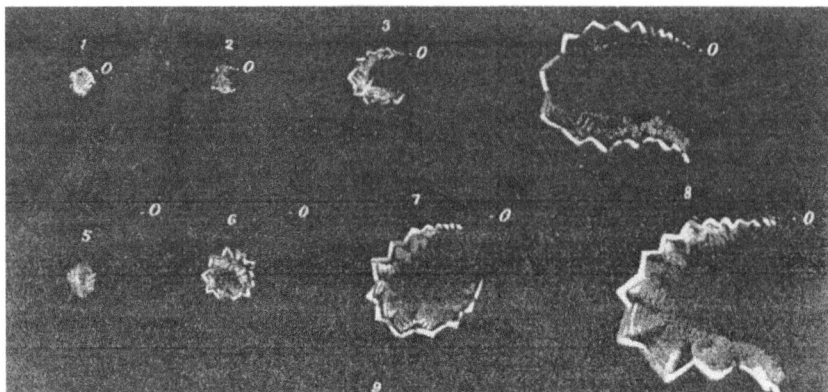

Figure 10.7. Drawing of fortification spectra experienced during a migraine aura by Dr. Hubert Airy. The numbers indicate the gradual enlargement of the zigzags and of the following scotoma. The tilted ovals show the fixation point in the visual field. Dr. Airy's father, the Astronomer Royal, communicated his son's findings to the Royal Society in 1870. (Airy H. On a distinct form of transient hemianopsia. *Philosophical Transactions of the Royal Society*. 1870; 160: 247–264)

color and the black-white sensitive neurons with concentric fields in areas V1 and V2. David Hubel, working with Margaret Livingstone, found that the color-sensitive cells were present in clumps ("blobs") among the columns of orientation-selective neurons (Figure 10.5).[18] Although these cells, like the geniculate neurons, had concentric receptive fields on the retina, the overlap of the fields had the potential to pinpoint stimulus locations—as Hubel recognized (Figure 10.9).

If the orientation-selective simple and complex cells are not providing higher order neurons with information about the shapes of objects, what else might they be doing? One suggestion is that they, together with cortical neurons in area V4, are monitoring the level of brightness in the visual field.[16] This assessment was critical because the amount of excitation exhibited by a color-sensitive neuron depends not only on the wavelength of the light but on the intensity (luminosity) of the latter. Thus, without a correction for intensity, weak neuronal excitation could be due either to relatively bright light having a wavelength well to one side of the maximum or to dim light in the optimal wavelength. By making corrections for luminosity in different parts of the visual field, the cells in V4 could enable "color constancy" to be maintained regardless of brightness or dimness.[19] Another possible function for the simple and complex cells in V1 and V2 is the detection of movement in the visual field, so that the gaze could then be directed to the site in question for a more accurate appraisal. If either, or both, of these functions are correct, then the linear arrangement of the receptive fields can be seen as the most economical "packing" system in the primary visual cortex.

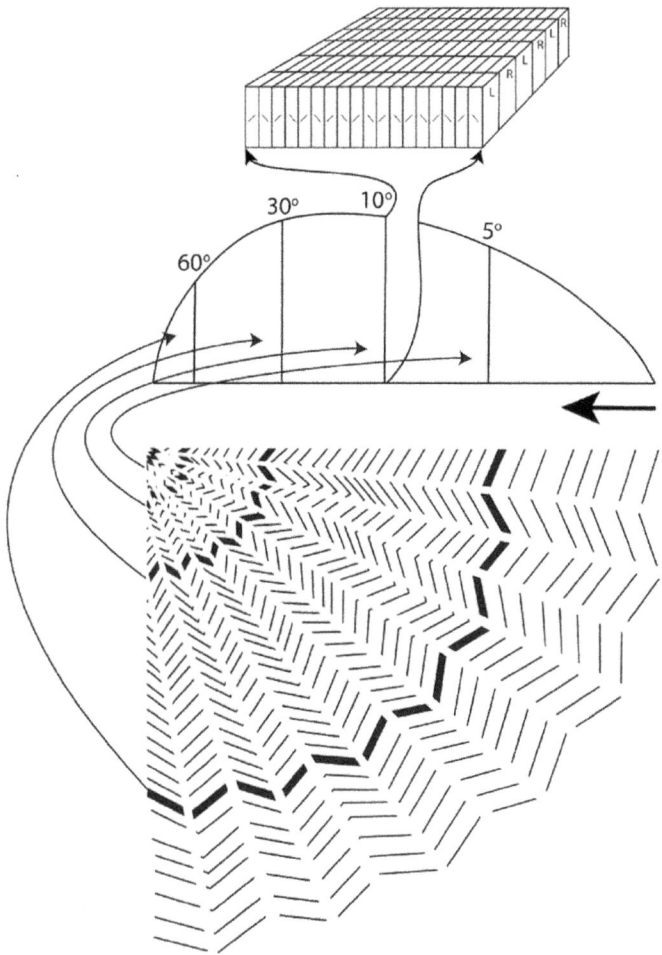

Figure 10.8. Diagram suggesting how zigzag lines of migraine aura are produced. A traveling wave beginning at the back of the brain (most posterior part of the occipital lobe) moves slowly forward over upper V1 (*arrow*, middle section of figure), exciting orientation-selective cells with ever-enlarging receptive fields and thereby producing zigzags in visual field (bottom of figure; four typical zigzag bands shown). The top of the figure shows the arrangement of the cortical columns for the orientation selective cells. Any linear feature in the visual field will fail to excite the orientation selective cells if it is orthogonal to the zigzags or lies along their junctions. (From: McComas A, Upton A. The migraine aura: a problem for vision theory. *Reviews in Biomedical Engineering.* 2016; 44: 347–355 with permission of Begel House, publishers)

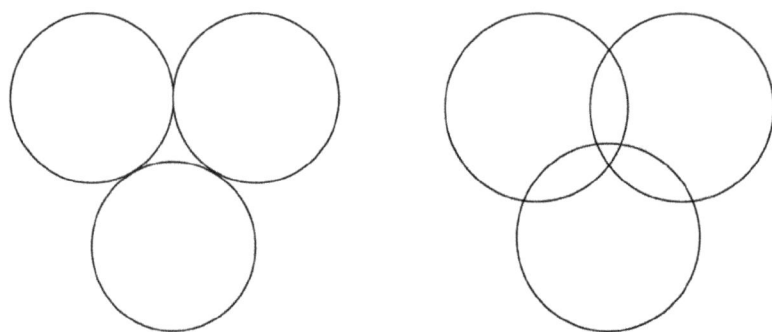

Figure 10.9. Localization by overlapping retinal fields. In (*a*) the three fields are separate and only three loci can therefore be recognized. In (*b*) the slightly larger fields overlap, the various combinations yielding seven possible loci.

Object Recognition

Regardless of whether visual processing is carried out in a hierarchical or parallel manner, how is it that an object can be recognized by the brain? A key observation here was the discovery, by Edmund Rolls and his colleagues at Oxford University, of neurons in the temporal lobe of the monkey that responded best to pictures of faces.[20] Subsequent studies by Doris T'sao at Harvard provided further details, showing that some cells in the inferotemporal cortex responded best to parts of faces while other cells recognized entire faces, in some cases regardless of the angle of the head. Intriguingly, the investigators also detected cells that fired most strongly when presented with pictures of faces that the animal could not possibly have encountered before.[21] It seems that the visual system, once it has become familiar with certain faces, is capable of constructing novel ones. Indeed, it is possible that these imaginary faces are among those that the brain makes use of in dreaming.

Helmholtz and Learning by Unconscious Inferences

This account leaves open the question as to how the cells in the inferior temporal lobe become responsive to faces to begin with. And, for that matter, how do other cells in the same lobe come to show selectivity for cars, tables, boats, trees, and animals, for example?[22] At a still more fundamental level, how does the brain, during development, acquire the ability to interpret certain features in the visual field as those of an object?

One of the first to consider this problem was Hermann von Helmholtz (1821–1894). Helmholtz (Box 10.2), a true genius who seemed able to make

Box 10.2 Hermann von Helmholtz (1821–1894)

Hermann Helmholtz was born in Potsdam, in what was then Prussia, in 1821; his father was a headmaster who would take a lifelong interest in his son's career while Hermann's mother was a direct descendant of the American Quaker, William Penn. On finishing his schooling, Helmholtz entered the military medical school in Berlin where he came under the influence of the great neurophysiologist Johannes Müller. After graduating, Helmholtz started his obligatory military service as an army physician. Fortunately his remarkable scientific gifts attracted attention, and he was allowed time to pursue his own interests in a hut set aside for this purpose. Helmholtz's first major publication reflected not only his brilliance but his grasp of disparate fields. He proposed that, regardless of the activity taking place, there was always conservation of energy—energy could not be created or lost, though it could take a different form (very often, heat). In 1849 Helmholtz was appointed associate professor of physiology at the University of Königsberg, and it was there that

Figure B10.2. Portrait of Helmholtz as a young man. (Wellcome Collection CC BY)

> he measured impulse conduction velocity in the frog sciatic nerve and then in human motor nerve fibers. The importance of this work extended beyond the values obtained—like his study of muscle contraction, it took away some of the mystery of the body.
>
> While at Königsberg, Helmholtz invented the ophthalmoscope, building one quickly himself from odds and ends and using his observations to illustrate lectures to the medical students. Further research, some of it in Bonn and Heidelberg, led to a trichromatic theory of color vision, an understanding of the visual mechanisms involved in depth perception, and the suggestion that perception relied on "unconscious inferences" acquired in early life. Helmholtz was also interested in acoustics, carrying out experiments on human volunteers and showing that complex sounds could be broken down into harmonics.
>
> A brilliant mathematician, Helmholtz was one of the pioneers of non-Euclidean geometry, and he also undertook the first analysis of liquid flow in vortices. Indeed, it was his wide range of interests that led him to accept his final academic position, that of professor of physics at the University of Berlin in 1871. Renowned throughout the scientific world and the mentor of young men who would themselves achieve fame, Helmholtz was elevated to the Prussian nobility in 1883.

a major contribution to any problem that interested him, considered cognition throughout his life. As a young man, working as an associate professor of physiology in Konigsberg, he was the first to measure the velocity of the nerve impulse. Two years later he proposed a trichromatic theory of color vision and invented the ophthalmoscope. In 1867, then at the University of Heidelberg, he developed the concept of sensory learning by unconscious inferences. By 1885 he was professor of physics in Berlin, a position he held for 14 years and, on top of his other duties, took it upon himself to give science lectures to the public. It was then that he returned to the subject of unconscious inferences in the acquisition of sensory judgments. As an example, Helmholtz emphasized the importance of movement, either of the body or simply of the eyes, as the same part of the visual field was viewed from different angles and thereby recognized as including an object. There was an analogy with touch sensation in that, by running a finger over surfaces, the brain could appreciate something as an object possessing a certain size and shape. Learning, then, was empirical in that the brain of the newborn had no in-built neural assemblies for recognizing objects; instead, the ability came from exploration, from trial and error.[23]

Studies of Previously Blind and Deaf Individuals

As it turned out, Helmholtz was correct in emphasizing movement as a cue to object recognition by the developing visual system, though not necessarily in the way he had imagined. Rather than active movement on the part of the observer being required, "passive" movement within the visual field of a stationary observer was also sufficient. This very important conclusion came from the study of blind persons who had their sight restored. A few such individuals were reported over the years, including one studied in depth by Oliver Sacks,[24] and a common feature was their visual confusion, such that they were unable to make sense of what they were looking at. This type of study took on a new dimension when a large number of blind children were treated surgically.

These Indian children had been blind from birth, either from congenital cataracts or from intrauterine infection by trypanosomes. The young patients were old enough to be able to describe what they were seeing after their surgery, and it emerged that objects came to be identified as such because certain parts of the visual field moved together.[25] This was not entirely surprising given that the same principle could be demonstrated in normally sighted persons who, confronted with a confusing picture of black spots on a white background, were immediately able to discover a hidden Dalmatian hound once certain spots moved in unison. The ability of the visual system to create outlines was also apparent in the illusion of the Kanizsa triangle (Figure 10.10), in which the

Figure 10.10. The Kanizsa triangle illusion, in which the brain completes the interrupted sides.

brain inferred the existence of the three sides. As the neuroscientist Dale Purves pointed out, there are many examples of unconscious inferences in our normal perception of the world, and especially in the judgments of distance and size.[26]

The developing brain is able to acquire sensory discrimination, then, by the unconscious recognition of commonalities—in the case of object recognition, by detecting linked movements in the visual field. This is not the only clue that the visual system uses, however, for differences in color between an object and surroundings must also be important. In the developing infant there comes the ability to correlate what is seen with what is felt with the hands or brought to the mouth.

In the case of the auditory system, the brain also appears to work by associations, though in time rather than in space. The analogy with the restoration of sight is, of course, the return of hearing to the deaf, especially by the increasing use of cochlear implants. The basic operation of the implant is the electrical stimulation of fibers in the auditory nerve in response to sounds. The latter are picked up by a microphone worn by the deaf person; the sounds undergo an analogue-to-digital conversion and, after digital processing, result in a sequence of electrical pulses delivered through electrodes implanted in the cochlea. As there are only 22 electrodes available in an implant, compared with the 30,000 hair cells that normally excite the auditory nerve fibers, the loss of information is enormous. Nevertheless, by detecting repeated patterns of impulses, preferably while viewing the movements of a speaker's lips, it becomes possible for the auditory cortex of the previously deaf person to ultimately recognize speech and even music, albeit in a highly distorted mode.

But while accepting that the conscious and unconscious inferences are being used higher up the visual pathway, is there anything more to be said about the functions of the primary visual area (V1)?

Enlarging the Role of the Primary Visual Cortex

If the parallel processing scheme is correct, then V1 begins object recognition by detecting linear features in the visual field. If, however, the orientation selectivity of the simple and complex cells is a misread clue as to their function, then the case for distributed processing is weaker, and the main function of V1 (and V2, V3) would then be to transfer largely unprocessed information to the inferior temporal lobe. Major processing would then take place in the temporal lobe, resulting in the creation of groups of neurons that would respond specifically to a particular face or other object. Such groups of neurons would correspond to Konorski's "gnostic units," discussed in the previous chapter, and their formation would be enhanced by repetition or by the accompaniment of strong emotion.

Among the various output connections of the facial gnostic units would be those to the hippocampus, where other features of the person (name, clothing, dwelling place, actions, etc.) would be combined with the face cells into higher order gnostic units, the so-called concept cells.

Recent studies have suggested, however, that V1 may have another, and very important, role in addition to the passing of largely unprocessed information on to the temporal and parietal lobes. Indeed, it seems likely that V1, assisted by inputs from the higher order visual areas, participates in the final "read out" of the visual field. This possibility has been raised by several types of observation; historically, the first of these is the "masking" experiment.

Backward Masking and "Time-Chunks"

The investigation of backward masking began with the psychologists in the mid-19th century. It was found that a weak sensory stimulus could not be perceived if a stronger one was given shortly after. This was as true for electric or tactile stimuli applied to the skin as for brief flashes of light or pulses of sound. Experiments such as these showed that the brain's interpretation of an early signal was affected by the processing of a later one, provided that the interval between the two was no greater than 30 ms to 100 ms (depending on the respective stimulus strengths). An especially striking example of this principle was provided by the experimental psychologist Ron Efron, who flashed a red light quickly followed by a green one, causing the subject to see neither red nor green but yellow![27] Efron's psychologist predecessors in the previous century had been able to produce similar effects by spinning colored wheels.

Experiments such as these strongly suggested that the brain, most likely the cortex, analyzed information conveyed by the senses in "chunks" of time, each lasting 30 to 100 ms or so.[28] In a laboratory, the onset of a time-chunk was determined by delivery of a stimulus. In the natural world, however, an internally generated event would suffice, such as one of the constantly occurring shifts of gaze between different parts of the visual field ("saccadic" eye movements). The function of area V5 in the middle temporal lobe might then be to "smooth out" the chunks, so that sensation appeared continuous—just as happened in the cinema, where successive movie frames were projected faster than the critical flicker fusion frequency.[29] More dramatic examples of the ability of the visual cortex to create the illusion of movement came from the stage, in the sleight-of-hand tricks performed by magicians.[30]

Notwithstanding the uncertainty over hierarchical/parallel schemes, evidence that some of the visual processing during the time-chunk was done in V1 itself came from several sources. First, Vahé Amassian (Box 10.3) and his

Box 10.3 Vahé Amassian (1924–2013)

With his Oxford accent, elegant turn of phrase, and impeccable manners, Vahé Amassian appeared the quintessential English gentleman. In fact he had been born in Paris to Armenian parents and had gone on to spend most of his working life in the United States. And, rather than Oxford, it was at Cambridge that he had studied for a medical degree. Like a number of other would-be physicians, he had become infatuated by neurophysiology, in his case largely due to the influence of Edgar (Lord) Adrian, the Nobel Laureate and head of the Cambridge Laboratory of Physiology during the war years.

Rather than approach Adrian and stay in Cambridge, however, Amassian emigrated to the United States, taking a position with Arnold Towe at the University of Washington in Oregon. There, Americanized to the point of smoking cigars and driving a large car, Amassian began his study of the mammalian cortex in earnest, initially with microelectrode recordings—then in their infancy— and later with stimulation of the pyramidal tract. After a move

Figure B10.3. Vahé Amassian during a conference in Warsaw, 2009. (Author's photograph)

> to the Albert Einstein College of Medicine in New York, Amassian subsequently became professor and chair of the Department of Physiology at the SUNY Downstate Medical Center where, from 1972 onward, he was to remain for the rest of his working life.
>
> It was in the latter part of his career that Amassian recognized the opportunities provided by magnetic stimulation for exploring the human brain and proceeded to work out its mode of operation. With colleagues in New York (notably Roger Cracco) and in Queen Square, London (John Rothwell), Amassian exploited TMS to the full and was probably the first to employ it as a means of temporarily inactivating localized areas of cortex. A stream of original papers followed, including those that analyzed the time-course of the steps involved in seeing, recognizing, and naming a letter of the alphabet.
>
> Hugely knowledgeable, a polished speaker at meetings, and an infectiously enthusiastic experimenter and colleague, Vahé Amassian was, in defiance to his small stature, an academic giant.

group at New York University had subjects read, and then say, a letter of the alphabet that was flashed on a screen.[31] Up until 140 ms after the stimulus, V1 was clearly engaged since transcranial magnetic stimulation (TMS) delivered to V1 within this time abolished the perception; after 140 ms, there was no effect. Since it took 40 ms for a visual stimulus to start any neural activity in V1, then V1 must have been active for 100 ms or so—the same duration as a time-chunk. This conclusion was consistent with the results of recording neural activity from V1; there was a small early wave, and this was followed by a much larger one.[32] Microelectrode recordings from single neurons confirmed the presence of the later activity, and both TMS studies and latency measurements at different points in the visual pathway provided an explanation for it—V1 received a projection back from the higher visual areas.[33,34]

Thus, rather than simply passing information on to higher visual areas so that objects and other features in the visual field could be recognized, V1 shared in the final integrated perception (Figure 10.11). Importantly, as Efron's color-flash experiments demonstrated, the perception took into account *all* the information that had accumulated in V1 during the time-chunk.

Reconsideration of the Function of Gnostic Units

If a perception depends on the back projection to the corresponding sensory receiving area in the brain, what is the role of the gnostic unit(s) at the head of

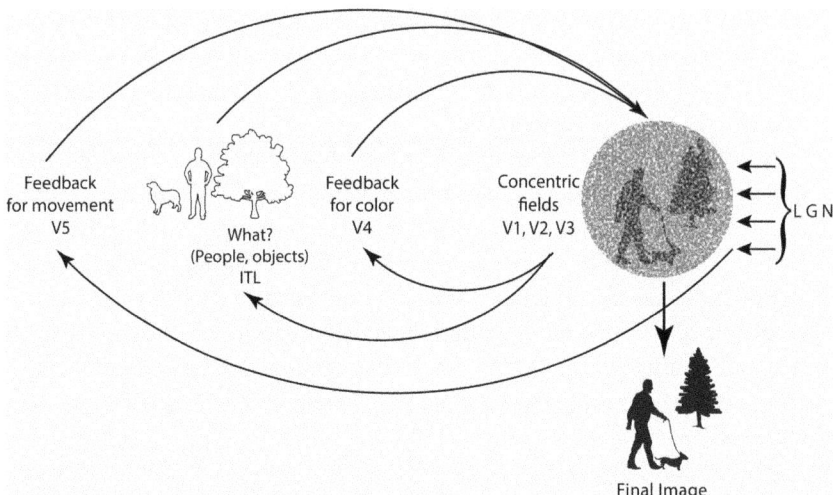

Figure 10.11. Alternative scheme for visual pathways, in which the final image is formed by feedback to area V1, with content having been recognized in the inferotemporal area (*ITL*).

the sensory pathway? The answer is that the gnostic units are essential if there is to be any "meaning" attached to the response of the primary receiving area. Remember that congenitally blind children, on being given vision through eye surgery, were initially unable to comprehend what they were looking at—their visual world consisted of a collage of colors. Only when parts of the visual field moved in unison was it possible for them to discern shapes of objects and people. This, then, is what the gnostic units in the visual and other sensory pathways do—they make sense of what is experienced. In the visual pathway it is the gnostic units in the inferotemporal cortex that respond to outlines and, through emphasis on salient features, provide a coded caricature of a person or object. (Black and white is evidently sufficient for this, as shown by the responses of temporal lobe neurons to uncolored images presented on a computer screen.) The visual perception, then, is the result of activity in the gnostic units in the inferotemporal lobe (recognition of shapes and feature enhancement) and in the primary visual cortex (for detail and coloration), brought together at the end of a time-chunk. We can say, then, that the sensory experience has three parts: (a) *presentation* of impulse activity to the primary cortical receiving area; (b) *recognition* of shapes in the inferotemporal cortex, and (c) *summation* of back-projected neural activity with that present in the primary cortical area for color, detail, and place.

The combination of activity in gnostic units and in primary cortical receiving areas would be a very economical way of processing information by the brain. In the case of vision, for example, there would be no need for color analysis by a separate brain area—the function of V4 may simply be to provide a back-projection to the color-sensitive neurons in V1 for the "summation" phase that concludes a time-chunk. Similarly, the sensation of movement in the visual field could simply be due to back-projection from V5 to the "complex" cells in V1 and V2— the neurons discovered by Hubel and Wiesel that responded to movement of a critically oriented linear stimulus.

Returning to the gnostic units in a sensory pathway: apart from being necessary for the interpretation of what has been viewed (or touched or heard), the gnostic units have other functions. They serve as memories for persons, objects, words, sounds, tastes, and smells. And, by interacting with other gnostic units, they are responsible for mental or physical activity, as described earlier.

Conclusion

At the start of the chapter, vision—or rather visual processing—was stated to be a complex affair, and so it has proved. Uncertainty persists over the significance of some of Hubel and Wiesel findings, and over the presumption of a distributed analysis, operating in parallel, followed by recombination ("binding"). Related to this, the possibilities raised by back-projections to the primary receiving areas is taken up again in Chapter 13. Regardless of the interpretation of the results, however, the many experiments and the wide variety of approaches have enabled a huge amount of pertinent information to be gathered. And there are some conclusions that appear inescapable—that the immediate transactions in the primary visual cortex take place in columns of cells, for example, and that there are cells in the inferior temporal lobe that have learned to recognize faces and objects. Time-chunking, too, seems to be established, as does a late role for the primary visual cortex. It is more than a good start. We are well on our way to understanding visual mechanisms—except for that wretched "hard" problem!

After all the complex investigations and observations in vision, it is time to consider the results of an experiment that, though simple in concept and technique, has had a huge effect on our understanding of the relationship between neural activity and consciousness.

Notes and References

1. Hubel DH. *Eye, brain and vision*. New York, NY: Freeman, 1988.
2. Kuffler's experience in Australia with Eccles and Katz is described in McComas AJ. *Galvani's spark: the story of the nerve impulse*. New York, NY: Oxford University Press, 2011.
3. Kuffler SW. Neurons in the retina: organization, inhibition and excitatory problems. *Cold Spring Harbor Symposia on Quantitative Biology*. 1952; 17: 281–292.
4. De Monasterio FM, Gouras P, Tolhurst DJ. Concealed colour opponency in ganglion cells of the rhesus monkey retina. *Journal of Physiology*. 1975; 251: 217–229.
5. Hubel DH, Wiesel TN. Receptive fields, binocular interaction and functional architecture in the cat's visual cortex. *Journal of Physiology*. 1962; 160: 106–154.
6. Hubel DH, Wiesel TN. Receptive fields and functional architecture of monkey striate cortex. *Journal of Physiology*. 1968; 195: 215–243.
7. Mountcastle VB. Modality and topographic properties of single neurons of cat's somatic sensory cortex. *Journal of Neurophysiology*. 1957; 20(4): 408–434.
8. Lorente de Nó R. The cerebral cortex: architecture, intracortical connections and motor projections. In: Fulton JF, ed. *Physiology of the nervous system*. London, England: Oxford University Press, 1938:291–346.
9. The best description of the model is given in Note 1. See also: Ts'o DY, Zarella M, Burkitt G. Whither the hypercolumn? *Journal of Physiology*. 2009: 587: 2791–2805.
10. Zeki S. The Ferrier Lecture 1995. Behind the seen: the functional specialization of the brain in space and time. *Philosophical Transactions of the Royal Society B*. 2005; 360: 1145–1183.
11. Zihl J, von Cramon D, Mai N. Selective disturbance of movement vision after bilateral brain damage. *Brain*. 1983; 106: 313–340.
12. Singer W. Synchronization of cortical activity and its putative role in information processing and learning. *Annual Review of Physiology*. 1993; 55: 349–374.
13. Crick F, Koch C. Towards a neurobiological theory of consciousness. *Seminars in the Neurosciences*. 1990; 2: 263–275.
14. Concerns of the author.
15. Zeki S. The distribution of wavelength and orientation selective cells in different areas of monkey visual cortex. *Proceedings of the Royal Society of London B*. 1983; 217: 449–470.
16. McComas A, Upton A. The migraine aura: a problem for vision theory. *Reviews in Biomedical Engineering*. 2016; 44: 347–355.
17. Richards W. The fortification illusion of migraines. *Scientific American*. 1971; 224: 88–96.
18. Livingstone M, Hubel DH. Anatomy of a color system in the primate visual cortex. *Journal of Neuroscience*. 1984; 4: 309–356.
19. The majority of cells in V4 respond to white light or are maximally sensitive to more than one wavelength (Schein SJ, Desimone R. Spectral properties of V4 neurons in the macaque. *Journal of Neuroscience*. 1990; 10(10): 3369–3389). These properties would be consistent with V4 serving as a brightness monitor to effect color constancy. However, such a mechanism would not be necessary if there was back-projection to V1 (see further discussion in Chapter 12).

20. Perrett DI, Rolls ET, Caan W. Visual neurones responsive to faces in the monkey temporal cortex. *Experimental Brain Research*. 1982; 47: 329–342.
21. Meyers EM, Borzello M, Freiwald WA, Tsao D. Intelligent information loss: the coding of facial identity, head pose, and non-face information in the macaque face patch system. *Journal of Neuroscience*. 2015; 35: 7069–7081.
22. Majaj NJ, Hong H, Solomon EA, DiCarlo JJ. Simple weighted sums of inferior temporal neuronal firing rates accurately predict human core object recognition performance. *Journal of Neuroscience*. 2015; 35: 13402–13418.
23. Helmholtz H von. *Treatise on physiological optics*. Translated from the 3rd German edition (1910), Southall JPC, ed. Washington, DC: Optical Society of America.
24. Sacks O. *An anthropologist on Mars: seven paradoxical tales*. New York, NY: Knopf, 1995.
25. Sinha P. Once blind and now they see. *Scientific American*. 2013; 309: 48–55.
26. Purves D, Morgenstern Y, Wojtach WT. Perception and reality: why a wholly empirical paradigm is needed to understand vision. *Frontiers in Systems Neuroscience*. 2015; 9: Article 156.
27. Efron R. The duration of the present. *Annals of the New York Academy of Sciences*. 1967; 138: 713–729.
28. The time for a time-chunk is very much shorter than the 600 ms allotted to sensory processing by the brain by Libet (see next chapter) and casts doubt on the validity of the latter's experimental approach.
29. The critical flicker fusion frequency for human subjects is given as 60 to 90 Hz, but there are a number of factors that can raise or lower it. For cinema and TV, the effective frequencies are 24 Hz and 25 to 30 Hz, respectively.
30. The author experienced a striking example of the brain's ability for visual extrapolation while watching a baseball game. From his position in the stand, the baseball, after leaving the pitcher's hand and reaching the batter, consistently appeared to pass *behind* the latter's body on its way to the catcher.
31. Amassian VE, Cracco RQ, Maccabee PJ, Cracco JB, Rudell A, Eberle L. Suppression of visual perception by magnetic coil stimulation of human occipital cortex. *Electroencephalography and Clinical Neurophysiology*. 1989; 74: 458–462.
32. Clark VP, Fan S, Hillyard SA. Identification of early visual evoked potential generators by retinotopic and topographic analyses. *Human Brain Mapping*. 1994; 2: 170–187.
33. Buffalo EA, Fries P, Landman R, Liang H, Desimone R. A backward progression of attentional effects in the ventral stream. *Proceedings of the National Academy of Sciences*. 2010; 107: 361–365.
34. Pascual-Leone A, Walsh V. Fast backprojections from the motion to the primary visual area necessary for visual awareness. *Science*. 2001; 292: 510–512.

11
Benjamin Libet's Big Experiment

Development of Signal Averaging

Had he not become a physician, George Dawson (Box 11.1) could easily have been an engineer. As a young lad growing up in Manchester, England, he had a fascination with electrical and mechanical gadgetry—"a misspent youth" as he would laughingly refer to it. Later, soon after World War II, it was this fascination that led him to tackle one of the major problems in the study of the human nervous system, namely, the difficulty in detecting the very small electrical signals developed in response to stimulating the peripheral nerves of the arm or leg. Recording from muscles was not difficult—the hundreds of muscle fibers supplied by a single motoneuron acted as biological amplifiers for that nerve cell in the spinal cord. Indeed, long before the invention of the cathode ray oscilloscope, investigators had been able to record electrical impulse activity in muscles; later, muscle potentials were used to study reflex activity in the human spinal cord[1] and to establish the velocity of the impulses travelling in the motor nerve fibers.[2] Recording from the nerve fibers supplying the skin, muscle, and joints presented a greater challenge, however, the potentials being a thousand times smaller than those from muscle. The same problem bedeviled attempts to detect the responses in the brain to stimulation of the peripheral nerves. Not only were the evoked potentials of low amplitude—because of the scalp and skull under the recording electrodes—but they were mixed in with the much larger spontaneous EEG waves.

Dawson's initial approach to both situations was surprisingly simple.[3] All that had to be done was to make the trace on the oscilloscope screen very faint while superimposing tens of responses, all the time keeping the camera shutter open. Because the nerve or brain potential always occurred at the same time after the start of the trace, whereas the "noise" was generated randomly, it—the evoked response—could be seen against the fuzzy background of the noise. While this strategy worked well for the nerve responses, it was less satisfactory for the brain. A new approach was needed.

The idea came after Dawson had given a communication of his response superimposition work at a meeting of the Physiological Society in London. A member of the audience[4] made the suggestion that, rather than the cortical responses being superimposed, they should be added together and their average

Box 11.1 George Dawson (1912–1983)

George Duncan Dawson was born in Manchester, England. His father was a pathologist at the Royal Infirmary, and it was there that Dawson undertook medical studies, graduating in 1936. As a boy he enjoyed making electronic apparatus and was also good with machinery, so it was therefore not surprising that he spent a research fellowship designing and constructing an EEG machine for the leading neurosurgeon, Sir Geoffrey Jefferson. On the outbreak of World War II, Dawson joined the Royal Air Force but, having contracted tuberculosis, was invalided out. At his own request he convalesced in an epileptic "colony," so that he could continue his research on the EEG. At that time the discipline was in its infancy, and its few practitioners, including Grey Walter and Dawson, would meet periodically to compare progress. In 1944 Dawson joined an academic research unit at the Hospital for Nervous Diseases in Queen Square, London, where he was able to continue his studies on epilepsy, while at the same time embarking on a quest to detect very small

Figure B11.1. George Dawson in his laboratory, probably in the late 1950s. (Author unknown)

> bioelectric potentials (see main text). His first signal averager worked well but, because of advances in digital electronics, was never reproduced; it is now in the collection of the National Science Museum in London. Dawson also held a position in the Department of Physiology at University College London, where he was made professor in 1966. There, the door of his small laboratory was ever open to students and staff seeking the advice and practical help—which would be freely given by the small, cheerful man inside. In the summer he would often join the same students and staff in dinghy sailing. In addition to his signal averaging and EEG accomplishments, Dawson investigated the possibility of feedback control from the cerebral cortex on to ascending somatosensory pathways, using the rat as his model.
>
> George Dawson was respected and admired. It was felt by many that his breakthrough in detecting very small biological potentials deserved a Nobel Prize, and it is a sad reflection on the honors system that this gifted man was never elected to the Fellowship of the Royal Society.

taken. The logic behind the suggestion was that the background noise, being random, would be as often negative as positive in the recording system at any instant after the stimulus and would thus cancel itself out; the nerve-induced response, however, would consistently have the same polarity.[5]

But how to create the average of the brain potentials? Dawson, drawing on his knowledge of machinery and electronics, designed a machine for the task. A rotary switch with two banks of contacts was employed; each contact completed a circuit from the signal amplifier (connected to electrodes on the scalp) to one of 62 capacitors. With each revolution of the switch, following a stimulus to the nerve, the charge at any instant was fed to a corresponding capacitor. The stimulus would be repeated and the potentials developed at the same times after the stimulus would be added to the same capacitors. At the end of a number of repetitions the charges stored by all the capacitors would be read out in sequence—and so was born the science of signal averaging in neurophysiology.[6]

No sooner had Dawson established the validity and usefulness of signal averaging for biological potentials than another development occurred. Rather than using motor-driven capacitors to store charges, one could achieve the same end by converting the biological potentials into a series of pulse trains. Thus the number of pulses in each train was proportional to the amplitude of the response at a given moment; after repeated stimuli the number of pulses stored at each of the locations in the computer memory was averaged and then displayed as a voltage. The device performing these functions became known as the CAT (Computer of Average Transients) and was an invention of the Australian scientist—and concert pianist—Manfred Clynes. The CAT resembled an oscilloscope both in its modest size and in its appearance. Unlike the much larger LINC computers, which had already appeared in some laboratories, the CAT did not require any programming—all one had to do was press the right buttons. Further, it could do more than average synaptic potentials; it could also analyze trains of impulses generated by nerve or muscle cells. During the short time that these devices were commercially available, they would find a place in virtually every neurophysiological laboratory.

The Bereitschaftspotential (Readiness Potential)

Freiburg is one of Germany's many beautiful historic cities. Situated in the southwest corner of the country, close to the borders with France and Switzerland, it lies near the banks of the Rhine and, in the opposite direction, faces the mountains and the Black Forest. Rising high over the roofs of the old city is the spire of the Catholic cathedral, a spire that has dominated the skyline for the past 500 years. It was during a lunch in one of the city's fine restaurants, not far from the cathedral, that the idea came for a radical extension of the human cortical recording experiments.

At the time Hans Kornhuber (Figure 11.1) was the chief physician in neurology at the Freiburg university hospital; the head of the neurology department was the eminent clinician-scientist Richard Jung.[7] Kornhuber was born and raised in Königsberg, only to be interned for four years by the Soviets at the end of World War II. It was the latter experience that prompted him to switch from chemistry to medicine at college. Having qualified at the University of Heidelberg, Kornhuber obtained a higher degree for a thesis on visual and vestibular integration. Throughout his career he continued to have strong leanings to research and would later spend time in Vernon Mountcastle's laboratory at Johns Hopkins Hospital; it was there that he compared the responses of touch-receptor fibers cells in primates with the sensations evoked in humans by similar vibratory stimuli.

On that spring day in 1964, however, Kornhuber was with his doctoral student, Lüder Deecke, and as they sat in the garden of the restaurant, their thoughts turned to the possibility of finding an electrical potential in the human brain that might signal the "will" to make a movement. The solution that they came up with was elegant in its simplicity; it was to make a continuous tape recording of the brain activity (using EEG electrodes on the scalp) while a subject made tapping movements with his or her hand whenever he or she wished. The tape was then turned over in the recorder so that it could be played backwards; in this way the brain activity before each of the movements could be examined with the averaging mode of the laboratory's CAT. To the delight of the experimenters, a

Figure 11.1. Hans Kornhuber, 1928–2009. (Photograph by Jürgen C Aschoff. Wikimedia Commons CC BY-SA 4.0)

small potential *could* be seen; it was a negativity that started almost a second before the movement took place and gradually built up, becoming steeper in the last 0.5 seconds (Figure 11.2, top). The two young men gave the name *bereitschaftspotential* (readiness potential) to their discovery, and the paper reporting their findings would become a classic.[8,9]

Unbeknownst to Kornhuber and Deecke, during the very same period, a leading neuroscientist in Britain was making similar observations. William Grey Walter was the senior scientist at the Burden Neurological Institute in Bristol. Under Grey Walter's leadership, the Institute, which was set in the grounds of a large mental hospital, had been highly active in EEG research and in the use of intracerebral wire electrodes to better locate and treat functional disorders. A pioneer in robotics, Grey Walter was known to the public for his creation of autonomous "tortoises."[10] In 1964, however, the main interest in the laboratory was in recording potentials that might be associated with readiness to move. Using a "barrier grid" storage device (similar to a radar tube), Grey Walter was able to show a surface negativity developing over the fronto-vertex region as, following a warning signal (a click), the subject waited for a second signal (a flash) commanding him or her to press a button.[11] The increasing negativity was referred to as the "contingent negative variation" or the "expectancy wave" and was essentially the same as Kornhuber and Deecke's readiness potential. Possibly because the latter's experiment was the simpler of the two, in that no warning and command signals were involved, it was the readiness potential that attracted the greater attention in the neuroscience community. Since both studies used EEG electrodes on the scalp for the recordings, the next step was to identify more

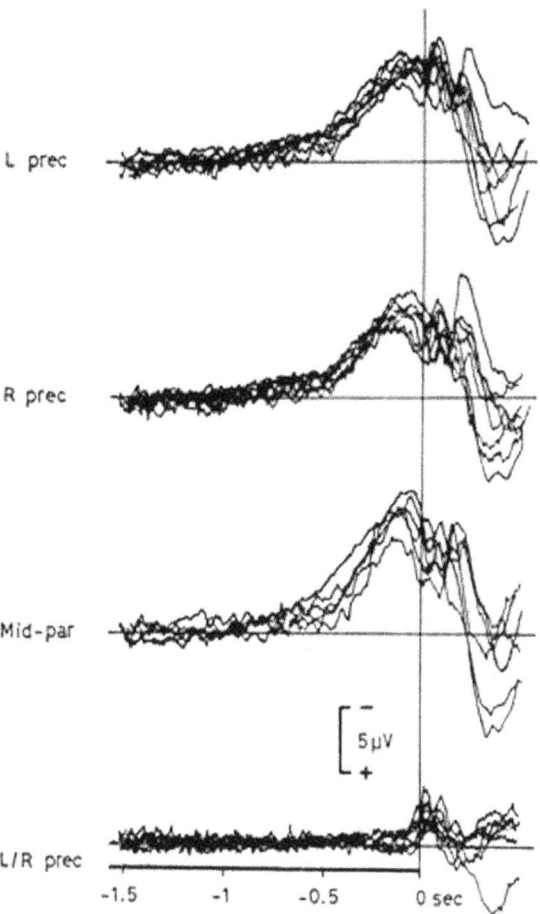

Figure 11.2. Kornhuber and Deecke's recording of the bereitshaftspotential following "voluntary" hand tapping. There are five to eight averaged traces for each site on the scalp, and they show the wide distribution of the potential (*prec*, precentral; *mid-par*, mid-parietal). (English Wikipedia; author Lüder Deecke)

precisely which cortical area(s) were generating the slow negativity. Various types of observation in animals and humans suggested that there was involvement of the supplementary motor area—that region of cortex on the medial side of the hemisphere immediately in front of the motor cortex itself.[12]

Enter Benjamin Libet

The fact that a signal could be recorded from the brain, one that apparently reflected the "will" to move, attracted enormous interest—understandably so. One of those impressed was Benjamin Libet (Box 11.2). At the time Libet was at the University of California San Francisco, and, while there, he had the opportunity to make experimental observations in the operating theatre through his neurosurgical friend and colleague, Bertram Feinstein. A skilled neurophysiologist, Libet planned those experiments carefully. He concentrated on the somatosensory cortex and, when the latter was exposed during surgery, stimulated it electrically and noted the feelings reported by the patient. That, in itself, was nothing new—Wilder Penfield and his associates in Montreal had been doing the same thing extensively since the 1930s. However, Libet had gone a step further by noting the change in stimulus parameters that caused a previously undetected stimulus train to be felt (on the opposite side of the body). If the intensities (i.e., the currents) of the stimuli were kept constant and sufficiently weak, he found that a stimulus train had to last at least 500 ms to be felt. From this he deduced that 500 ms was the minimum time required for cortical processing to generate a conscious sensation.[13] This finding and conclusion were surprising for it had been shown long before (using dissected nerve fibers in medical students!) that a single impulse in a single nerve fiber coming from the skin could probably be detected.[14]

The second reason for surprise was that a stimulus, regardless of its nature, seems to be recognized the moment it is delivered, rather than half a second later. Libet countered the first objection by pointing out that, no matter how brief it was, a sensory stimulus evoked responses in the cortex that lasted hundreds of milliseconds. The second objection—the "immediacy" of the perceived stimulus—required a more ingenious explanation. Libet suggested that, just as the somatosensory cortex was able to refer sensations to a particular point on the opposite side of the body ("spatial reference"), so it could refer sensations to an earlier moment —the time when impulse activity was first initiated in the cortex following the stimulus ("temporal reference").[15] These were important conclusions and inevitably became the subjects of debate following their publication. But Libet was soon to deliver a greater surprise.

Box 11.2 Benjamin Libet (1916–2007)

Like Jerry Lettvin (Box 9.1), Benjamin Libet was born in Chicago and the son of immigrants, his having fled with their parents from persecution in the Ukraine in the early 1900s. Again, like other immigrant families, his struggled during the Depression. Nevertheless, Benjamin—through scholarships—was able to attend the University of Chicago where he came under the influence of the neurophysiologist Ralph Gerard. Rather than proceeding to medical school like Lettvin, Libet was immediately attracted to the life of a neurophysiologist and went on to earn a PhD under Gerard's supervision, studying EEG rhythms and steady potentials in the excised frog brain. After a period as an instructor at the Albany Medical College, New York, Libet was called to serve in World War II, where he conducted research on survival clothing for airmen who had crashed in water. After the war he returned to the University of Chicago, working on the spreading depression of Leão. There were also summer visits to the Marine Biological Laboratory at Wood's Hole, where he

Figure B11.2. Benjamin Libet at a conference on consciousness in Montreal, 1997. Though he does not look it, Libet was in his 80s. (Author's photograph)

found marked ATPase activity in preparations of membrane from the squid giant axon—narrowly missing the discovery of the Na^+-K^+ pump. Next came a move to the University of California, San Francisco, and a sabbatical in Canberra with Jack Eccles. It was there that he was introduced to the sympathetic ganglion preparation by Eccles' daughter, Rose; it was a preparation her father had used in the 1930s to investigate synaptic transmission. The ganglion became the focus of Libet's research for the next years in San Francisco; among his many findings was "long-term enhancement," a form of potentiation that was mediated by dopamine and could persist for hours. It was while he was successfully engaged in synaptic mechanisms that Libet was invited to tackle something very different—an investigation of neural activity in the exposed brains of patients undergoing surgery (see main text). Because of his remarkable finding, that a cortical potential developed well in advance of a conscious desire, Libet remained a major figure in international neurophysiology for the remainder of his long life, receiving awards and participating in international symposia.

It is rightly said that one good idea, pursued well, is all that is required for a successful career. A striking example was James Watson's fixation on DNA for study and the proposal of the double helix by himself and Crick soon afterwards. Another, taking an example from the brain, would be the recording of the readiness potential (bereitschaftspotential) by Kornhuber and Deecke. All four investigators in these two examples were young men. When Libet made his most important discovery he was well past normal retirement age, though with his full head of jet-black hair he looked much younger. And, once again, just as with Kornhuber and Deecke's discovery of the readiness potential, Libet's experiment was simplicity itself.

Libet had his subjects press a key whenever they wished, while at the same time recording the readiness potential, just as in the studies of Kornhuber and Deecke. In the interpretation of such an experiment it was reasonable to assume, as Eccles had done,[16] that the "will" to move coincided with the start of the potential. Libet's critical addition to the experiment was to test this assumption. He did so by having the subjects watch an oscilloscope screen on which a spot of light circled; they then had to report the position of the spot at the moment they became aware of the desire to move (Figure 11.3). Since the velocity of the traveling spot was known, the time of appearance of the intention could be simply calculated. The results were both convincing and unexpected—Libet found the readiness potential was already well developed (by some 250 ms) before the desire appeared[17]; thus a decision only entered consciousness when the underlying

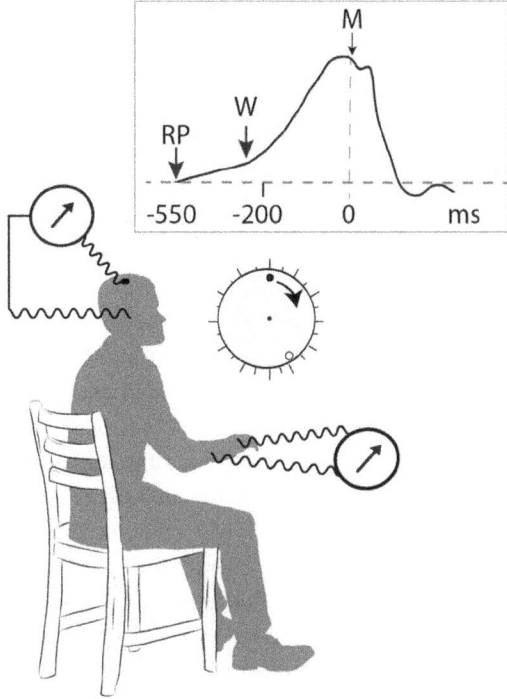

Figure 11.3. Libet's free will experiment. The subject observes a circling dot on the oscilloscope screen and notes its position when he or she is aware of the intention to move. Meanwhile simultaneous recordings are made of electrical activity in the brain and in the (tapping) muscles on the back of the forearm. *RP*, onset of readiness potential; *W*, awareness of will to move; *M*, muscle activation.

neural activity was already far advanced (Figure 11.3, top). Rather than the mind controlling the brain—thought by thought—it was the other way round, and "free will," seemingly so self-evident, was an illusion. Was it any surprise that there was consternation within the scientific and philosophical communities!

Commentary

In view of their importance for an understanding of brain mechanisms, Libet's results had to be considered further. While there could be no doubt of their creator's ingenuity and of the directness of his approach to the study of consciousness in human subjects, were the results as clear-cut as they seemed? And could they be confirmed by other investigators?

Cortical Processing Time

The minimum processing time required for a sensation, in Libet's experiments employing direct cortical stimulation, was stated to be half a second or so. Was it possible that this was the time required for the successive cortical responses, evoked by the train of shocks, to sum to some critical value needed for conscious awareness? If it were, then one could consider the minimum time to have been artificially prolonged. It had long been known, for example, that in various parts of the nervous system repetitive stimuli could cause excitatory postsynaptic potentials to overlap and build up until the depolarizations reached the critical firing levels for the neurons involved. Libet was well aware of this potential objection and made additional observations. He recorded the electrical responses in the cortex following repetitive stimulus pulses to a neighboring area and found that the responses remained separate and of the same amplitude.

Libet's further results might have ended the argument had not a new method for stimulating the cortex been devised. It was a method that did not require the surface of the brain to be exposed, was painless and free of risk, and could be applied to any sensory receiving area. This new approach was *transcranial magnetic stimulation* (TMS). When the veteran New York neurophysiologist Vahé Amassian[18] delivered TMS pulses to a site anterior to the motor cortex, some of the subjects were able to feel them as a touch or a tingle on the opposite side of the body; provided the stimulus was large enough, there was no need to repeat it—a single pulse was sufficient to elicit a conscious sensation.[19] These last observations would seem to repudiate Libet's conclusion that ongoing impulse activity lasting half a second or more was necessary for consciousness. Yet another argument against Libet's long processing time came from the results of choice reaction time experiments. In such experiments, the subject had to discriminate between two or more sensory stimuli (implying conscious activity) and then make an appropriate motor response as quickly as possible. The typical reaction times for this sort of experiment were in the 200 to 250 ms range, far shorter than Libet's times for conscious intent.

Free Will Experiment

It was Libet's second experiment, the one that appeared to destroy the concept of free will that attracted the most attention, however. Libet's observations were repeated in a number of laboratories with variations in the methodology, both in the nature of the task and in the conscious and instrumental reporting. Despite the modifications, the results confirmed Libet's findings. In one study, for example, Soon and colleagues had their subject recognize an alphabetical letter on the screen in front of the observer, rather than the position of a moving dot. Again, instead of recording the electrical activity of the brain, functional magnetic resonance imaging (fMRI) was employed. Lastly, while making observations on the supplementary motor area (SMA), as Libet had done, they also included other regions of the brain in their study. Soon et al.'s results proved even more remarkable than Libet's, for the fMRI appeared to detect consistent activity in the frontal and cingulate cortices as long as 7 to 10 seconds prior to movement.[20]

In view of their importance, the design and interpretation of Libet's free will experiment were examined very thoroughly, and it was not long before a number of objections or concerns were raised.[21] Though there are others, there were five that can be considered here.

Visual Processing Time
Reporting the position of a dot (or identifying a letter) did not take into account the time required for visual processing. Libet's explanation, already given earlier, was that the brain automatically referred the moment of conscious awareness to the onset of electrical activity in the brain, in this case in the primary visual cortex.

SMA as the "Wrong" Cortical Area
Because of its well-investigated, and apparently dedicated, function in programming the motor cortex, the SMA was an unlikely site for the "will" to arise. Although activity here might be a precursor of the will, increasing the possibility of such an intention arising, it was not the will itself. A more likely site for the will was the parietal lobe, where the presence of multiple sensory inputs suggested more complex functioning and where a variety of observations, including the readiness potential, indicated the presence of "cognitive" areas related to forthcoming movements.[22] This objection would not destroy Libet's hypothesis, however, but merely modify it by moving it to another part of the brain.

The Ability to Cancel an Intended Movement
This was a more serious objection, and one that Libet had considered. He pointed out that, even after the intent to move had appeared, the subject could still cancel the movement—that is, he or she could say "no." Libet's explanation was that this was an example of an intrinsically conscious thought acting on the brain (i.e., true free will). If this were not so, the time between the onset of a readiness potential and the will to cancel must be considerably shorter than the interval between the start of the readiness potential and the unobstructed movement—thereby throwing doubt on the significance of such long intervals. However, an

alternative explanation was that the decision to cancel had already been made by unconscious brain mechanisms and then appeared later as a conscious act.

Simplicity of Task
Pressing a button was a very simple action. How could the experimental findings made with such a basic task possibly be related to very much more rapid movements that appear to be under conscious control? Examples of the latter would include the choice of words in rapid conversation and the playing of a musical instrument at a fast tempo. Libet was aware of this last objection too, being a talented musician (singer) himself, and pointed out that performers concentrate on the sound to be produced rather than on the details of its execution—leaving the latter to occur automatically.[23]

The "Rules of the Game"
The Libet experiment assumed that the participating region of brain was idle before the intention to move appeared. In fact, there would have been ongoing neural activity informing the subject that he or she was to make a certain movement whenever he or she wished—what might be termed "the rules of the game." Further, Libet's experiment required the subject to watch the oscilloscope screen for the first appearance of the circling dot, a mental activity likely to be associated with a recordable expectancy wave, as in Grey Walter's experiments (see previous discussion). Given these additional operations on the part of the brain, how could one be sure that it was the entire readiness potential that gave rise to the intention to move, rather than a much later component?

While the Libet experiment was regarded by many as the long-awaited solution to the mind-body problem, the uncertainties provided sufficient ammunition for the "dualists" to persist in their belief in free will. The policy taken in this book is that, despite the arguments presented here and despite some corrections of timing, Libet was essentially correct. In the next two chapters, ones that attempt a synthesis of all the previous material, the issue of free will is taken up again. Can there be anything more important?

Notes and References

1. Hoffmann P. Über die Beziehungen der Sehnenreflexe sur willkürlichen Bewegung und sum Tonus. *Zeitschrift fur Biologie.* 1918; 68: 351–370.
2. Hodes R, Larrabee MG, German W. The human electromyogram in response to nerve stimulation and the conduction velocities of motor axons. Studies on normal and on injured peripheral nerves. *Archives of Neurology and Psychiatry.* 1948; 60: 340–365. Although this paper is usually given precedence, Helmholtz had measured impulse conduction velocities in human motor nerve fibers a century earlier, using a mirror galvanometer.

3. Dawson GD, Scott JW. The recording of nerve action potentials through skin in man. *Journal of Neurology, Neurosurgery and Psychiatry*. 1949; 12: 259–267.
4. Surprisingly, not a fellow neurophysiologist but a gastric physiologist, Dr. J. N. Hunt.
5. Dawson pointed out that the same principle had been used in the previous century in attempts to detect an influence of the moon on atmospheric pressure.
6. Dawson GD. Cerebral responses to electrical stimulation of peripheral nerves in man. *Electroencephalography and Clinical Neurophysiology*. 1954; 6: 65–84.
7. See Figure 3.3.
8. Kornhuber HH, Deecke L. Hirnpotentialänderungen bei Willkürbewegungen und passive Bewegungen des Menschen: Bereitschaftspotential und reafferente Potentiale. *Pflügers Archiv für die gesamte Physiologie des Menschen und der Tier*. 1965; 284: 1–17.
9. Deecke L, Grözinger B, Kornhuber HH. Voluntary finger movement in man: cerebral potentials and theory. *Biological Cybernetics*. 1976; 23: 99–119.
10. See Chapter 14 for more on Grey Walter.
11. Walter WG, Cooper R, Aldridge VG, McCallum WC, Winter AL. Contingent negative variation: an electric sign of sensorimotor association and expectancy in the human brain. *Nature*. 1964; 203: 380–384.
12. Fried I, Mukamel R, Kreiman G. Internally generated preactivation of single neurons in human medial frontal cortex predicts volition. *Neuron*. 2011; 69: 548–562.
13. Libet B, Alberts WW, Wright EW, Delattre L, Levin G, Feinstein B. Production of threshold levels of conscious sensation by electrical stimulation of human somatosensory cortex. *Journal of Neurophysiology*. 1964; 27: 546–578.
14. Hensel H, Bowman KA. Afferent impulses in cutaneous nerves of human subjects. *Journal of Neurophysiology*. 1960; 23: 564–578.
15. Libet B, Wright EW, Feinstein B, Pearl D. Subjective referral of the timing for a conscious sensory experience: a functional role for the somatosensory specific projection system in man. *Brain*. 1979; 102: 193–224.
16. Eccles JC. In discussion. Cited in: *Mind time: the temporal factor in consciousness*, by Libet B. Cambridge, MA: Harvard University, 2004.
17. Libet B, Gleason CA, Wright EW, Pearl DK. Time of conscious intention to act in relation to onset of cerebral activity (readiness potential). The unconscious initiation of a freely voluntary act. *Brain*. 1983; 106: 623–642.
18. For more on Amassian, see Box 10.3.
19. Amassian VE, Somasunderinn M, Rothwell JC, Cracco JB, Macabee PJ, Day BL. Parasthesias are elicited by single pulse magnetic coil stimulation of motor cortex in susceptible humans. *Brain*. 1991; 114: 2505–2520.
20. Soon CS, Brass M, Heinze H-J, Haynes J-D. Unconscious determinants of free decisions in the human brain. *Nature Neuroscience*. 2008; 11: 543–545.
21. Klemm WR. Free will debates: simple experiments are not so simple. *Advances in Cognitive Psychology*. 2010; 6: 47–65
22. Anderson RA, Buneo CA. Intentional maps in posterior parietal cortex. *Annual Review of Neuroscience*. 2002; 25: 189–220.
23. This is certainly true, but the argument would not apply if one were to play an unfamiliar work requiring quick fingering.

12

More on Gnostic Units and Cortical Columns

We are now at a point where some of the pieces can be fitted together. First, however, the earlier conclusions bear repetition.

Recapitulation

(i) In animal species, including birds and perhaps even insects, degrees and types of consciousness are generated by nervous systems that vary widely in size and structure. (Chapter 4)

(ii) In the mammalian brain the cerebral cortex, working in conjunction with the thalamus, is involved in the generation of consciousness. (Chapters 6, 7)

(iii) Different parts of the mammalian cortex subserve different aspects of consciousness. (Chapter 6)

(iv) Part of the interaction between the two sites, cortex and thalamus, is through rhythmic electrical activity; the thalamus influences impulse generation in cortical cells. (Chapter 7)

(v) It is through its impulse discharges that the brain fulfills its various functions and, at the same time, gives rise to consciousness. (Chapters 8, 9)

(vi) The reticular activating system in the brain stem is especially important in controlling the level of excitability of the cortex. (Chapter 7)

(vii) In a sensory system, the mainstream view is that different aspects of the stimulus are processed simultaneously in different cortical areas and the results combined to yield the conscious perception ("parallel processing"). (Chapter 10)

(viii) The alternative is that the brain, through its sensory systems, learns about itself and its environment empirically, using assemblies of neurons to create "gnostic units" for categories and individual objects and persons. (Chapters 9, 10)

(ix) The alternative view also emphasizes the importance of back-projections to the primary sensory cortex. In the visual system, the back-projections are responsible for the perception of color and detail in the image and perhaps of movement too. (Chapter 10)

(x) In much of the mammalian cortex (the "neocortex") functional units consist of columns of neurons perpendicular to the surface of the brain. (Chapters 6, 10)

Types of Gnostic Unit

History and Description

As noted in Chapter 9, the term "gnostic unit," used to describe a neural assembly having knowledge (information), was the invention of Jerzy Konorski (1903–1973).[1] At the time Konorski was professor of neurophysiology at the Nencki Institute in Warsaw, Poland. He had trained in medicine and spent two years under Pavlov in St. Petersburg studying reflex behaviors. Prior to his Russian experience he had, with a fellow student, demonstrated the possibility of what is now known, through the work of Skinner, as operant conditioning.

Gnostic Units for Categories

Like Pavlov, Konorski regarded the different sensory systems as "analyzers." To a large extent, his concept of gnostic units came from his experience as a neurologist, during which time he observed patients who, usually as a result of strokes or brain injuries, had developed particular inabilities. For example, after a lesion involving part of a cortical visual area, it could be difficult or impossible for the affected person to recognize faces or objects or perhaps written words. Such selectivity implied that different cortical visual areas contained gnostic units coding for specific categories.

Figure 12.1, reproduced from Konorski's *Integrative Activity of the Brain*,[2] depicts some of the visual categories. In addition to human faces (row [d]) and signs and handwriting (rows [g], [h]), there are categories for small manipulable objects and larger objects, facial expressions, human figures and animals, and limb positions.

Just as he had done for the visual system, Konorski was able to use clinical observations to identify specific categories of recognition for the auditory system (the "acoustic analyzer"). These categories comprised nonverbal sounds, human voices, words, and melodies. Similarly, for the "somesthetic [body sensory]

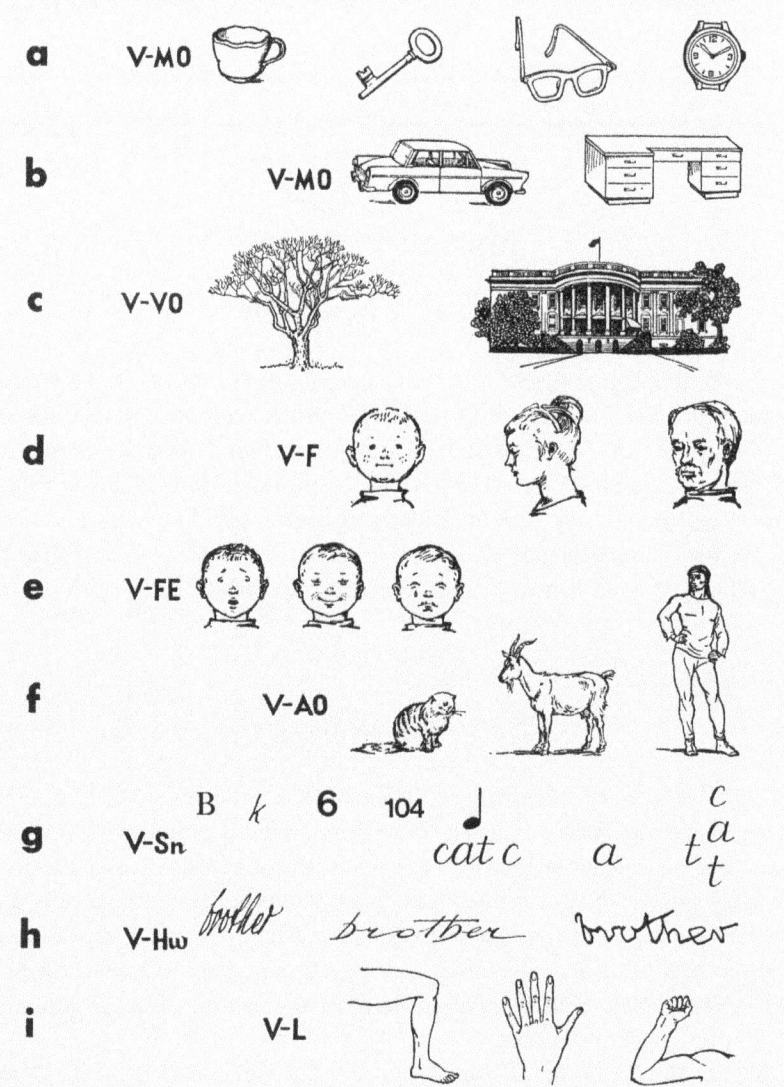

Figure 12.1. Jerzy Konorski's categories of gnostic unit. Each row (*a–i*) represents gnostic units responsive to a particular category of person, animal, object, etc. Within each row a gnostic unit will respond only to one individual. Neurological lesions may "wipe out" neurons for a whole category or, if smaller, neurons for an individual. (Reproduced with permission from Konorski J. *The integrative activity of the brain.* Chicago, IL: University of Chicago Press, 1967)

analyzer," there were categories for postures of the limbs and mouth and for different features of external objects (texture and taste).

Even in a normal subject there are readily available illustrations of mental behavior that are strongly supportive of the existence of neural assemblies for categories. One example involves the auditory analyzer and involves the search for a name. Suppose I want to telephone a former colleague in my university department. I know his first name is George, but I cannot recall his surname. Immediately my brain starts presenting me with the names of other Georges—rather like an old-fashioned telephone directory. Thus I become aware of "George Clooney . . . George Bush . . . George Gershwin . . . George Washington" and finally the name I want, "George Sweeney." Does this not suggest that my brain has a category for "George" and, almost certainly, similar categories for other first names? Another example of categorization involves faces. Not only does the category include real faces but the appearance of "faces" in, for example, the outlines of clouds or rock formations or in an abstract painting.

Gnostic Units for Individuals and Objects (Concept Cells)

Konorski identified the posterior region of the temporal lobe in humans as the site for facial gnostic units, knowing that ablation experiments in animals had shown the equivalent area in primates to be the inferotemporal cortex. Some 12 years after publication of Konorski's book, Edmund Rolls and his colleagues in Oxford did, in fact, show that individual cells in this part of the monkey brain responded to human and nonhuman primate faces.[3] In humans the right temporal lobe was subsequently found to be more concerned with faces than was the left, and it appeared that there was overlap between face-recognition and object-recognition neurons. In patients with temporal lobe damage, both types of discrimination could be lost, though in occasional cases only one was affected.[4]

As described in Chapter 9, a more complex type of gnostic unit, the "concept/grandmother cell," has been found in the hippocampus and in other areas of the brain (putamen, amygdala, and frontal cortex).[5] With increasing exploration of the human brain, it is likely that additional types of gnostic units will be detected in other cortical areas. One might, for example, eventually come across a concept cell for "red"—a neuron that would respond to the written or printed word, the spoken word, as well as the color itself. Similarly, after a prolonged search, one might discover a concept cell for "cat," in keeping with a line of thought by Peter Milner, a colleague of Donald Hebb in the Department of Psychology at McGill University: "I might recognize a cat by its smell, by the sound it makes, by sight, or by the combinations of these stimuli . . . A glimpse of ginger fur under a chair may elicit recognition of the cat."[6]

Not all concept cells arise from the natural workings of the brain. Some can be "consciously" formed as an aid to memory. As an example, I had a colleague, Mark, whose surname continually eluded me until I linked it to "diamonds." The thought of diamonds directed my mind to London's famous diamond market, Hatton Garden, and thence to the complete name—Mark Hatton. Most likely I had formed concept cells responsive to Mark's face, his written name and... diamonds.

Kinesthetic Gnostic Units

In addition to the gnostic fields for the special senses, Konorski postulated the presence of neural assemblies for various types of movement as part of a "kinesthetic analyzer." Here there were categories for identifying movements of the hand, body, and mouth, respectively, as well as for locomotion and the organization of motor behavior. The kinesthetic analyzer differed from the other sensory analyzers in that the transit units conveyed information not from exteroceptive receptors (eye, ear, body surface, taste buds, olfactory receptors) but from muscle, joint, and tendon receptors. Through the latter group the cerebral cortex was continuously informed as to the positions of the head, neck, trunk, limbs, and fingers. An audacious proposal of Konorski's, discussed later, was that the kinesthetic units, once formed, became capable of initiating movements themselves.

Although the concept of kinesthetic gnostic units, including their muscle-activating role, is attractive and is adopted here, it is likely that Konorski erred in envisaging the cerebellum acting as an intermediary between the receptors and the kinesthetic analyzer. Instead, two lines of evidence indicate that the kinesthetic analyzer is more likely to be found in the parietal lobe than the frontal lobe. One type of observation, initially made by Vernon Mountcastle (Box 6.3) in Baltimore, is that neurons in the parietal lobe fire before those in the frontal lobe when a movement is about to occur.[7] The second observation comes from researchers in Lyon, France, who stimulated the exposed cortex of human subjects prior to brain surgery. Electrical stimulation of the right inferior parietal lobe resulted in a strong desire to move the contralateral hand, arm, or foot, while stimulation of the corresponding area on the left produced the intention to move the lips and to talk. With strong stimulation the patients believed, erroneously, that they had actually made the movement. In contrast, stimulation of the premotor area in the frontal lobe yielded the reverse—movements occurred without the patient being aware of them.[8]

Another modification to Konorski's concept of a kinesthetic analyzer arises from later electrophysiological studies indicating that both vision and touch also provide important information about body posture and movement. A key feature of this analyzer, pointed out by Konorski, is that, although the kinesthetic gnostic units do not normally make us conscious of our posture, we can nevertheless immediately identify the position of any part of the body—as in touching or reaching, for example. As discussed in Chapter 6, it was this unconscious image of the body that Henry Head, the English neurologist, termed the "body schema" and that later research identified with neural activity in Brodmann's areas 5 and 7 in the parietal lobe. If the schema is lost or distorted, the consequences are severe, as noted earlier.[9]

Anatomical Locations of Gnostic Fields

Konorski related his analyzers, with their gnostic fields, to Korbinian Brodmann's classification of cortical areas, presented in Chapter 6. Konorski accepted the results of classical brain mapping in identifying visual function with the posterior parts of the occipital lobes (Brodmann areas 17, 18, and 19). However, as already noted, Konorski regarded these as merely transit areas while the visual gnostic fields were to be found more anteriorly, in the parietal lobe (Brodmann area 39) and temporal lobe (Brodmann area 37). The somesthetic analyzer lay in the anterior part of the parietal lobe (Brodmann areas 1, 2, 3, 5, anterior 7 and 43) and the acoustic analyzer in the temporal lobe (Brodmann areas 21, 22, 41,42); the kinesthetic analyzer was situated, wrongly, in the frontal lobe.

Cortical Columns and Gnostic Units

Cortical Columns as Gnostic Units

What type of neural structure might correspond to a gnostic unit? At the simplest level, electrophysiological recordings have shown that a single neuron could be regarded as a gnostic unit, in that it could respond selectively to the sight of a particular face, for example. This was shown to be true of the pyramidal cells in the hippocampus, where the concept cells (Chapter 9) were located. The hippocampus has a relatively simple architecture, however, and single cell recordings in other parts of the cortex have shown that neurons responsive to closely related stimuli lie in columns perpendicular to the cortical surface. Mountcastle showed this type of assembly to be a feature of the somatosensory area, and Hubel and Wiesel subsequently found it to be true for the visual cortical areas as well. Moreover, the electrophysiological findings were consistent with the microscopic evidence of radially oriented columns of cells and fibers found by Lorente de Nó. Later anatomical investigations, this time with the electron microscope, showed that the apical dendrites of the pyramidal cells in the columns became entangled and formed bundles as they approached the cortical surface.[10] All of these types of study point to the neuron columns being functional units, and therefore these could also be considered as Konorski's gnostic units. This is the approach taken here.

We conceive, then, of sensory information being fed into the cerebral cortex through the respective receiving areas and being passed through a hierarchy of analyzers in the form of cortical columns. At the highest level will be the column(s) specific for a particular face or object—these cells will fire and the face or object will be recognized by the conscious brain.

The Creation of a New Gnostic Unit

Suppose the face or object encountered by an individual is a novel one. Clearly, the information about it will move through the analyzers to the gnostic units that best fit the new image. We thus mistake a stranger for someone we know who has a close resemblance. But a closer look provides new information and we realize our error. At the same time this additional information results in the recruitment of a further cortical column, a neighbor that is also receiving information from the column that had previously provided the mistaken image.

At this point there is a conceptual problem, and it has to do with the numbers of second-generation cell columns activated. The difficulty lies in the fact that each of these columns is surrounded by others—in Figure 12.2 these are shown in a hexagonal array. If these second-order columns were now to become recruited by yet another similar but novel stimulus, then a ring of third-order columns would form around each of the six second-order columns. Figure 12.2 shows that, in this example, there would be an additional 12 columns that had not been previously excited, for a total of 19 columns. This geometrical progression would soon become unwieldy, even for a cortical area rich in columns. So how can this be prevented?

The most likely answer is that, during consciousness, the excitability of an uncommitted cortical column—a "neighbor"—is fluctuating from moment to moment and doing so independently of those columns around it. In an alert subject this independence of the cortical columns is reflected in the absence of any clear rhythms in the EEG and only fast low-amplitude activity. There is also the observation that, in single-cell recordings from unstimulated cortex in humans and animals, the spontaneous firings of neighboring cells are not usually linked together.

To continue the explanation, excitation will build up spontaneously within a column until it reaches the threshold for the relatively large pyramidal "burst"

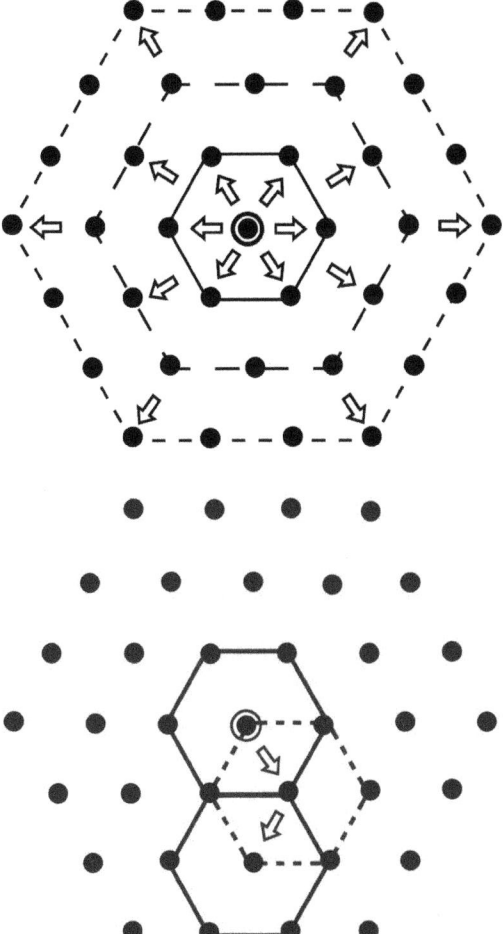

Figure 12.2. Alternative schemes for the spread of excitation between columns to form gnostic units; the two figures assume one is looking down on to the surface of the cortex and that each dot is a column (seen end-on). In the upper scheme, all the columns are available for recruitment and excitation spreads out in all directions (*arrows*) from the central column. In the lower scheme only one of the six columns surrounding the central column is responsive, as indicated by an arrow, and it will, in turn, only excite one of the six columns (also arrowed) in its own surround, and so on. After the third wave of excitation, all 37 columns will have been recruited in the upper scheme, compared with only 3 in the lower one.

cell(s) at the base (see later discussion). As this cell fires it will activate an inhibitory interneuron (basket cell or chandelier cell) through a collateral axonal branch (see Figure 12.3), making the column refractory for the duration of the inhibitory postsynaptic potentials (IPSPs). It is the large pyramidal cells in layer

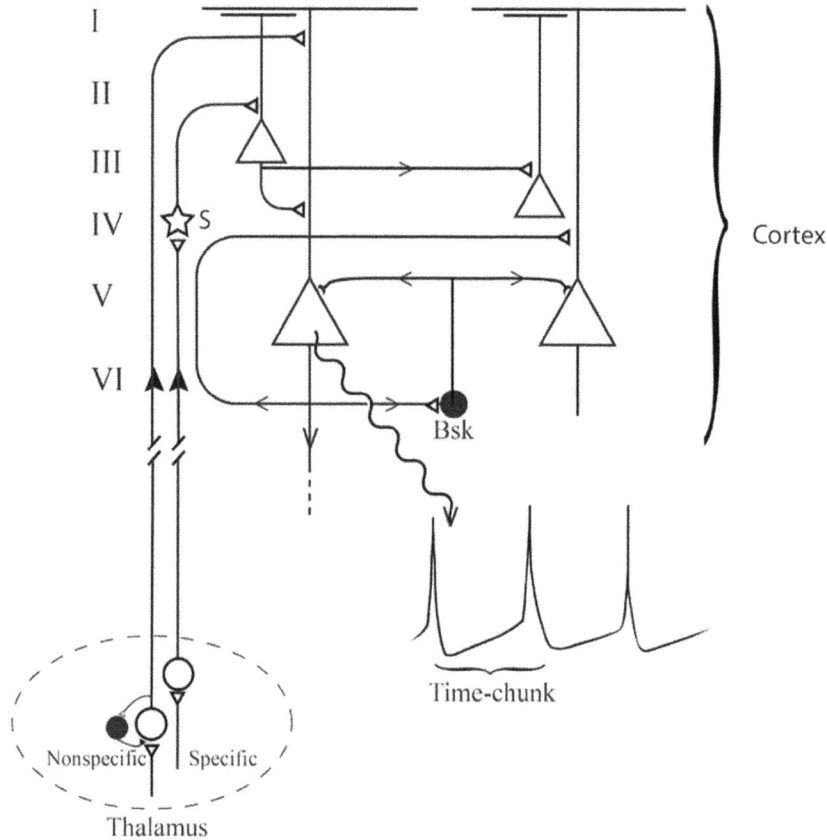

Figure 12.3. The flow of information through a cortical column. At bottom left, impulses from a specific thalamic nucleus excite a spiny cell (S) in cortical layer IV. The spiny cell then depolarizes the apical dendrites of the small pyramidal cells in layers II and III. The latter, in turn, depolarize the apical dendrite of the burst pyramid in layer V, generating an EPSP. If the latter is sufficiently large the burst cell may then fire; through a collateral branch of its axon a basket cell (*Bsk*) provides feedback inhibition (IPSPs) on to the layer V burst cell and its neighbors. As the inhibition wears off there is "rebound hyperexcitability" in the burst cells, which may sum with further EPSPs and cause them to fire again. One EPSP-IPSP cycle constitutes a time-chunk. In layers I and II the excitatory input from the nonspecific thalamic nuclei is directed to the apical dendrites of pyramidal cells and helps to control their responsiveness.

5, then, that control their respective columns. In the processing of a sensory message, only a column that has come out of the refractory period would be able to form a new gnostic unit in response to a sensory (or other) stimulus. Also helping to prevent the spread of activity through the cortex would be the surrounding gamma oscillation (IPSPs at 40 Hz) imposed by the thalamic projection nucleus, in keeping with the findings of Llinás, as described later.

Inherent in this scheme is the gradual spread of newly formed gnostic units away from the primary cortical receiving area as the brain accumulates sensory experiences. It is immediately around the receiving area that the "functional occupancy" of the cortical columns will be highest, while being lowest at the periphery of the sensory field.[11]

According to Konorski, the likelihood of a gnostic unit forming will be increased if there is an emotive component following the stimulus. This would be in keeping with our experience—the more dramatic a happening, the more likely it is to be remembered. In terms of brain circuitry, the limbic system would provide the emotive input to the cortical columns, including those forming the new gnostic unit. The larger the limbic input, the greater the number of columns likely to become gnostic units and the more vivid the recollected experience in the future. Once a unit has been incorporated in a chain of neural activity, it is probable that synaptic strengthening makes its commitment permanent, such that the cortical column cannot be taken over by another input. In situations where there is plasticity, it is the availability of "naive" columns that has made the creation of new gnostic units possible.

Why Cortex?

The question arises as to why, with the exclusion of pain and pleasure, sensations only appear to arise from activity involving cortical columns. Why, for example, is a subject with a posterior brain injury incapable of sensing light in the affected area of the visual field by virtue of the activity still present in the lateral geniculate nucleus? One answer is surely that there is something special about the structure, and therefore the function, of the cortical columns and, since the pyramidal cells are the predominant cell type, it is these that have to be considered. Leaving aside the characteristic shape of their cell bodies, the most striking feature of the pyramidal neurons is their apical dendrites—those long, tapering structures whose final horizontal branches form much of the outermost layer of the cortex.

A plausible hypothesis is that the prolongation of electrical activity in the apical dendrites (by virtue of prolonged excitatory postsynaptic potentials [EPSPs] and slowed impulse conduction in the fine terminal branches, reinforced by dendrodendritic connections) is one of the keys to consciousness. In other words, it is not so much whether a pyramidal cell body fires an impulse initially but whether, through a combination of incoming axodendritic and dendrodendritic activity, a sensory signal can persist as a depolarization in the dendritic layer for a critical time—the 30 to 100 ms corresponding to a "time-chunk."[12]

What Happens Within a Cortical Column?

The manner in which the cells within a column interact is critical for an understanding of consciousness and the many functions that the brain carries out. Not surprisingly, there have been a number of studies, the most comprehensive combining simultaneous intracellular recordings from two or more cells in the column(s) with identification of the same cells by means of injected dye—with the investigation carried out on brain slices. The interpretation of the data is not easy, for although the thickness of the neocortex appears to contain six layers of neurons and fibers, all but the two most superficial layers can be subdivided. Moreover, as Korbinian Brodmann and other anatomists recognized, there are variations in the six-layer structure from one region of the cortex to another. As if that were not enough of a problem, there are different types of neuron. For example, pyramidal neurons differ not only in their locations but in their sizes and distributions of their axons and dendrites. Moreover, some pyramidal neurons can fire steadily while others only in bursts (the "burst" cells in layer V). There are differences among the interneurons, too, including the presence of at least three types of inhibitory cells.

Despite the complexity of the task, however, significant progress has been made in unravelling the connections within a column.[13] Within a sensory pathway, for example, the excitatory input from the specific nuclei in the thalamus is directed to spiny interneurons in layer IV and thence to small pyramidal neurons in layers II and III and from these to the large pyramidal neurons in layer V (Figure 12.3). In contrast, the inputs from the nonspecific thalamic nuclei and from other areas of cortex terminate in layers I and II on the apical dendrites of the deep pyramidal cells. Of the nonpyramidal cells, the inhibitory interneurons clearly modulate the amount of excitation taking place within a column. One type, the basket cell, is of particular importance in that it is excited by, and feeds inhibition back on to, the layer V burst cells (see later discussion).

Cortical Columns and Consciousness

Relevant Experiments on Somatosensory Cortex

When the electrophysiological findings in the sensory cortex of an intact brain are combined with the columnar investigations described here, a picture emerges as to the generation of a conscious percept. Especially relevant are the studies of primary somatosensory cortex (S1) in monkeys by Cauller and Kulics at the Northeastern Universities of Ohio in Rootstown.[14] By varying the strength of a weak electric shock to the monkey's hand, Cauller and Kulics were able to correlate the awareness of the stimulus with the appearance of a late negative potential on the cortical surface; in keeping with a role in consciousness, they also noted that the same wave was depressed or absent during sleep or anesthesia. Following the example of Jasper and others, the authors observed how this cortical potential changed as the recording electrode penetrated deeper into the cortex; by this means they discovered that the potential was generated in the superficial cortical layers (I, II). After considering various possibilities, Cauller and Kulics attributed the late potential to feedback from "higher" somatosensory cortical areas, the returning fibers terminating on the apical dendrites of deep pyramidal cells.

By depolarizing the apical dendrites, then, the feedback can bring the layer V "burst" cells to the firing level. The long apical dendrites of the burst cells will already have received depolarizing inputs from the specific thalamocortical fibers (via the small pyramids in layers II and III; see Figure 12.3), and these inputs will help determine the strength of the burst cell firing.

The next step is the excitation of an inhibitory basket cell by the discharging layer V pyramid, resulting in feedback inhibition of the latter. Thus an EPSP and any superimposed impulses in the "burst" layer V cell are known to be followed by an IPSP.[15] As the IPSP wears off, the burst cell becomes hyperexcitable and will discharge again if sufficient depolarization has accumulated in its apical dendrite. Since, from masking experiments, the sensory cortex is known to operate in "chunks" of time, it is likely that each chunk corresponds to an EPSP-IPSP cycle, with the discharges of the burst pyramid generating consciousness of the stimulus (and an N1 response).

The burst cells dominate the column. Not only do they act as its time-keeper, but, through their long apical dendrites, they integrate the excitation within the column. The smaller, and far more numerous, superficial pyramids amplify the excitatory inputs to the column, before feeding them to the burst cells. They can collect and amplify while the burst cells are still recovering from their self-imposed IPSPs.

Though this "primary" column was the first to respond to a stimulus, its neighbors—which until then had been functioning independently of each other—may now join it and "beat" rhythmically. This is possible because the inhibitory interneuron fired by the axon collateral of the burst pyramid in the primary column projects to pyramids in other columns, evoking IPSPs and rebound excitation.

It is important to stress that the durations of the EPSP-IPSP sequences responsible for the time-chunks are not fixed. When cortical excitability is high, the sequences will be short enough to produce a gamma rhythm (40 Hz)—perhaps this is why, at times of acute stress, numerous thoughts can flash through the mind. When the sequences are long, as in the slow EEG rhythms characteristic of sleep, it may be their very infrequency that precludes consciousness.

One more feature of the column needs to be added. The burst cells, in addition to controlling the column, are known to project to the midbrain and superior colliculi. This projection could conceivably contribute to the "primordial" consciousness attributed to the brain stem by Damasio and others.[16]

In reviewing the evidence for this scheme, it could be objected that the use of an electric shock, as in the Cauller-Kulics experiments, is highly abnormal. Nevertheless there are many natural stimuli in the real world that are relatively abrupt—the constant saccadic movements of the eye (causing abrupt changes in retinal stimulation), a sudden noise, the foot striking the ground while walking, and so on. But suppose one closed the eyes and sat still in a quiet room—what then? Despite the absence of external stimulation, there would still be communication going on between cortical columns in the conscious brain. Like the back projections to a primary sensory area, the cortico-cortical axons would depolarize the pyramidal cell dendrites in layers I and II—increasing the discharge of the burst cells in layer V at the end of a time-chunk and contributing to consciousness.

The Critical Excitation Concept

Why, at any given moment, are some cortical columns contributing to consciousness and not others? The answer may lie in the intensity of the impulse activity in the small, and then the large, pyramids. Thus Steriade was able to demonstrate the great range of impulse frequencies exhibited by cortical cells, a range that extended from zero to hundreds of herz.[17] Further, there was a correlation between impulse frequency and state of arousal—the more aroused the animal, the higher the frequencies observed. Also pertinent were the findings of Rodrigo Quiroga and his colleagues on concept cells in the human hippocampus. They found that when the subject thought about the content of the picture being shown, the impulse firing rate quickened.[5]

Evidence from another direction are the observations on neurological patients; for example, in those with temporal lobe epilepsy, abnormal cortical impulse activity can create visual and auditory illusions or, in those who suffer with migraine, produce pain and sensory auras. In the case of migraine, subjects typically become aware of a flickering zig-zag pattern that slowly moves across the visual field and is most likely caused by excessive excitation of orientation-selective cells in the primary visual cortex of the opposite hemisphere. The point here is that the activity in these neurons, which is normally insufficient to produce a conscious image, is evidently capable of being elevated to the critical level during a migraine attack. That the headache of migraine is also due to excessive electrical activity in the cortex is indicated by the fact that, in a proportion of subjects, the symptom can be instantly abolished by a single pulse delivered from a magnetic stimulator[18]—probably through inhibition imposed by the cortical basket and chandelier cells. In some spontaneous pain syndromes the pain is so finely localized that it is likely to involve very few neuron columns in the cortex, on occasion perhaps only a single one. Indeed, experiments on functionally isolated slabs of mammalian cortex undertaken many years ago by the British neurophysiologist Delisle Burns suggested that the cortex may be organized in excitatory rings with surface circumferences of about 2 mm, a value little larger than the spread of the apical dendrite of a pyramidal cell.[19]

Numbers of Existing and Potential Gnostic Units

The next question concerns the number of neurons comprising a typical column, and this has been estimated as approximately 100; the great majority of these cells are pyramidal neurons (so termed because of the shapes of their cell bodies).[20] Since there are roughly 20 billion neurons in the human cerebral cortex, the maximum number of cortical columns would be rather less than 200 million. A proportion of these would be Konorski's "transit units" in the primary sensory receiving areas, but there would still be tens of millions of columns available as potential gnostic units in the association areas.

Tens of millions is an astonishingly large number; indeed, several lines of evidence indicate that the human cerebral cortex is more than capable of handling the amount of information required for the activities of daily living. First, there is the ability of human subjects to function normally, or almost so, after losing substantial parts of frontal cortex; this was true not only of Phineas Gage but of patients who underwent resection of prefrontal cortex for tumors or who had prefrontal leucotomies for behavioral disorders. Indeed, if performed early in life, it has been shown that a total hemisphere can be removed with remarkably little consequence, as in the case of a three-year-old girl described by the pediatric neurologist Karen Pape.[21] Also relevant are those hydrocephalic patients, a minority, who can function at a high level despite having only a thin rim of cortex surrounding greatly enlarged lateral ventricles.[11]

A second line of evidence pointing to a surplus of potential gnostic units for the needs of everyday life comes from the remarkable abilities of some individuals—the "savants"—to remember extraordinary strings of numbers or every word in entire books. Finally, there is suggestive evidence from the electrophysiologists themselves who, in their recordings from human and animal brains, find that only a small minority of cells can be influenced by different types of stimulus; that such cells are present is known from their spontaneous impulse firing. As indicated earlier, the surplus of cell columns is greatest in the frontal and prefrontal cortical areas and least in the posterior regions of the brain. Accordingly we can say that, in terms of their employment as gnostic units, there is a low "functional occupancy" of columns anteriorly and a higher one posteriorly.[11]

In sum, it is because of the huge numbers—millions— of cortical columns available that there are more than enough gnostic units to represent every item in a person's experience—every family member and acquaintance, every household gadget, every make of car on the road, every type of tree, flower, bird, and animal, as well as every word and phrase in that person's vocabulary. Similarly the cortex will host kinesthetic units that, acting in conjunction with other neural assemblies (basal ganglia, cerebellum), will code for every physical activity undertaken by the individual. There will also be kinesthetic units serving as "start," "stop," and "continue" commands. (There may even be kinesthetic [or, rather, "cognitive"] units that promote thinking and, as in the writing of a book on consciousness, thinking about thinking!)

Since cortical columns have formed a large part of this chapter and have had such significance attached to them, it is appropriate to finish this section with Figure 12.4, which shows a tiny piece of human cortex.

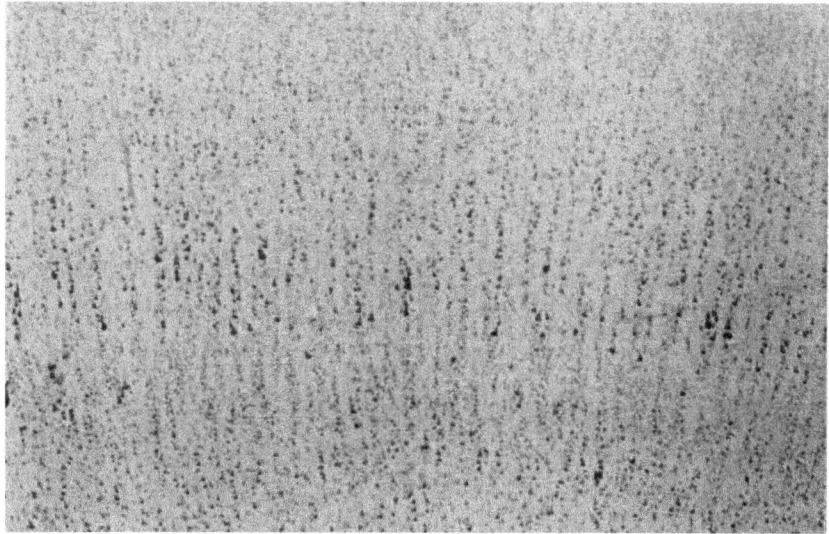

Figure 12.4. Nissl-stained section of human temporal cortex showing columns (vertical or near-vertical rows of cell bodies, most obvious if the page is tilted downwards; the thickness of the cortex is approximately 2.5 mm). (Courtesy of Dr. Sandra Witelson)

Creation Time and Processing Errors

Another question concerning the gnostic units concerns the time needed for their creation. As hinted earlier, it may require many repetitions of a stimulus pattern before a cortical column is committed to forming a unit, and this will be especially true if there is little or no associated emotional drive. Such a situation might be that of a new member joining a social club or of a student attending a new class, for only with repeated meetings is it possible for the newcomer to recognize the various individuals. At the other extreme there is evidence that, in certain circumstances, the visual analyzer can produce a new gnostic unit within seconds. A good example here would be the puzzle in which one searches for a hidden object or face within a deliberately confused picture. Once the hidden object is recognized, however, it can be detected immediately on the next presentation of the picture—presumably because a gnostic unit was formed at the time of initial recognition. Or, in another scenario, perhaps there was a man on the bus wearing distinctive clothing—a tartan waistcoat, say—who would be spotted instantly were he encountered two hours later in a restaurant.

Another aspect of visual gnostic units is the tendency of the analyzer to create novel ones in the absence of experienced combinations of features. Such units can be regarded as the results of processing errors—for example, if a unit is formed from the respective stimuli supplied by the nose of Person A, the eyes of Person B, and the hair of Person C. The occurrence of neurons responsive to such novel faces was noted in the inferotemporal cortex of monkeys tested by Doris Tsao and her colleagues at Harvard.[22] In human subjects these novel faces may appear in dreaming or in the hallucinations of patients with the visual deprivation characteristic of the Charles Bonnet syndrome.[23]

How Might the Gnostic Units Interact?

Suppose I go to visit my colleague Adrian in his university office (Figure 12.5). Adrian will become aware of my footsteps and of the perfunctory knock on the door through the acoustic analyzer in his temporal lobe. By means of connecting fibers from the temporal to the parietal areas, the auditory gnostic units excite kinesthetic units that will, in turn, activate the motor systems in the precentral cortex that cause him to swivel his chair round. Through the visual analyzer in his temporal lobe several classes of visual gnostic units now become active—one class for human figures, another for faces, and a third for facial expressions. All of these, working in conjunction, transmit information to his hippocampus where they excite the "concept cells" previously created for a familiar person—in this case, for me. (These "McComas" concept cells would also have fired if, on other

Figure 12.5. Gnostic units in action.

occasions, he had simply seen my printed name or heard my distinctive voice.) Next, through association fibers linking the visual (figure in doorway) and kinesthetic (stand up) gnostic fields in the temporal and parietal lobes, respectively, the motor apparatus in Adrian's precentral cortex is excited again. As before, this involves the supplementary and premotor areas, as well as the motor cortex itself, and will be associated with activity in the basal ganglia and cerebellum. Gracious person that he is, the ensuing muscle contractions enable him to smile and rise from his chair to greet me.

Gnostic Units and the Sense of Self

From infancy onward an ever-increasing amount of information is processed by the sensory analyzers, resulting in the creation of a large library of gnostic units. Such units are most likely to be formed if the same sensory information is repeated, and this by virtue of the Hebbian synapse principle ("cells that fire together, wire together"). Gnostic units will also be created if the information has a strong emotional content or is sufficiently unusual, as in the case of the man with the tartan waistcoat in the earlier example. A large fraction of the gnostic units will code for people, objects, and events in the external world. But there will be other gnostic units formed that have to deal with the person whose brain is carrying out all the sensory processing. In many instances these personal gnostic units have been, and continue to be, created by associations between the kinesthetic and the other sensory analyzers. Thus, whenever a kinesthetic analyzer prompts the motor apparatus into making a movement, there will be a new sensory input. It is likely that such associations begin before birth, and they will certainly be created in infants whenever objects are brought to the mouth, stimulating the exquisitely sensitive touch receptors in the lips. In another year or two, as the infant attempts to walk, there will be associations between the movement of the legs and the feeling of pressure on the soles.

The sensory-motor associations—and the creation and strengthening of gnostic units for self—continue throughout life. For example, if I now look down I see my fingers as they continue word processing, simultaneously exciting the touch receptors in my finger pads. Or if I go to a lumber store and choose a piece of wood by running my hands over its surface there will be the association of gnostic units for movement (the kinesthetic units) with those of touch (somesthetic units) and vision (observing the hands). It is through such combinations that gnostic units for "self" come to be formed. Moreover, it is a neural concept of self that is being continually reinforced, not least by activities that we pay little attention to, such as rubbing, scratching, or fidgeting. Adding to this self-concept or "schema" are gnostic units formed by sensory inputs from the viscera. Perhaps the best evidence for the existence of the body schema, and for its location in the parietal lobe, comes from observations on neurological patients with spatial agnosia—those individuals who, because of stroke, injury, or tumor, are no longer aware of the opposite sides of their bodies. As described in Chapter 6, presumptive evidence for the cortical location of the body schema is also to be found in a small minority of subjects with migraine, in whom disturbing illusions of body shape may occur prior to, or during, the characteristic headache.

Yet another class of gnostic units are those coding for thoughts. In the absence of any movement or ongoing sensory stimulation, it is possible for gnostic units to self-activate or to be driven by other units and for the result to be a thought. For example, the associated firing of gnostic units for my car, the railway station, and a previous look at the clock could lead to the thought of driving to the station in a hurry. Very often the thought is accompanied by an inner voice, the result of an association with auditory gnostic units.

Gnostic Units and "Voluntary" Movements

The key role of the kinesthetic analyzer in bringing about movements has already been alluded to. Once again the example of word processing is useful, this time in illustrating how this role might come about. A beginner, confronted with the keyboard, will develop the neural linkages between the visual system and the motor apparatus that will enable a chosen key to be pressed. With experience, however, there will be certain sequences of letters that will be used over and over again, as the same word is repeatedly typed. The corresponding pattern of finger movements around the keyboard will then become represented by a kinesthetic gnostic unit for the typing of that particular word. The information used in forming this unit will have come from receptors in the muscles, joints, and tendons that flex and extend the fingers, together with touch receptors in the finger pads. However, once the kinesthetic gnostic unit has been created by these sensory inputs, it becomes capable of functioning independently of them—the gnostic unit, activated by the sound or sight of the word, acts on the motor system to produce the typed word.

The principle of building complex kinesthetic units from simpler ones would apply to the acquisition of any manual skill. The principle would also be involved in the development of speech in an infant. In this case, information from sensory receptors in the lips, tongue, vocal cords, and chest—and also from the acoustic analyzer—would create simple gnostic units for sounds, and then, as speech develops, more complex units for words and ultimately for sentences. Although the results of introspection may be fallible, the concept of building up ever more complex kinesthetic units from simpler ones receives support from the playing of a musical instrument or the reciting of poetry. If in either case there is an interruption, it is much easier for the performer to go back and start the whole passage over again than to pick up the narrative at the moment of the break.

Recently there have been intriguing, if unusual, examples of the creation of kinesthetic gnostic units by patients with paralysis of all four limbs (quadriplegia). These remarkable studies have shown that it is possible for a patient to learn how to control a robotic arm by "thinking" of the movements required.[24] The associated impulse patterns are detected by an extensive array of miniature wire electrodes that has been placed in the motor cortex. The impulse patterns, in turn, undergo analogue-to-digital conversion and feed computer programs written for activating motors in the robotic arm. Unlike the normal situation, in which kinesthetic units depend heavily on inputs from muscles, joints, tendons (and skin), in the quadriplegic patients the only input for the creation of kinesthetic gnostic units comes from visual observation of the arm.

Now consider another type of "voluntary" movement or, rather, a series of movements in a fully mobile person. This time I walk into the kitchen, select a banana from a bowl of assorted fruit, peel it, and begin to eat it. What could be more voluntary than that? The gnostic unit system is quite capable of providing an explanation for this sequence of actions, however. In this example, it is the sight of the bowl of fruit that activates a corresponding gnostic unit (or units) in my visual cortex—I have, after all, seen many bowls of fruit in my lifetime. This gnostic unit has fiber connections to others, including those coding for the taste of a banana; these last units, in turn, activate the kinesthetic units involved in reaching down, grasping, and then peeling this choice of fruit. The important point here is that, although other sensory analyzers can achieve the same effects, it is vision, through its enormous battery of gnostic units, that is the most common instigator of "voluntary" movements. Further, the activation of the units and their kinesthetic partners intrudes into consciousness, such that there is a mental image of the action about to be executed. As Konorski pointed out, this last property—that of an imagined action preceding the actual one—was identified many years ago by the great pioneering psychologist William James (1842–1910; Figure 12.6), on the basis of his introspection.

In addition to the imagined act, there is one other aspect of this last example of voluntary movement that needs consideration and it concerns options. Not

Figure 12.6. William James (1842–1910). A Harvard medical graduate who intended to practice physiology but instead became the founding father of American psychology, teaching the subject at his alma mater. Like his brother Henry, the novelist, James was a prolific writer. (Photograph courtesy of Wikimedia Commons)

every time that I see a bowl of fruit do I stride across and remove a banana; indeed, there would be many embarrassments were this to happen. This lack of consistency is due to the operation of other gnostic units, including those for staying put and for the social milieu of the moment. Perhaps the most important factor in determining whether the gnostic units induce an action is the strength of the emotive drive—in this case, the presence or absence of hunger—operating through the limbic system.

Impulse Activity in the Gnostic Units: Role of the Thalamus

The activity of the gnostic units will be affected by those factors controlling the overall excitability of the cerebral cortex. In an earlier chapter, the reticular activating system in the brain stem was identified as having an especially important role, partly through neural connections through the nonspecific thalamic nuclei and partly through the routing of monoaminergic (and cholinergic) fibers to the cortex. The hypothalamus, too, was shown to have a generalized effect, in this case by the secretion of orexins. In addition, cortical excitability was affected by sensory inputs having relatively direct access, as in the case of vision. We open our eyes as we wake up, and, if tired, sleep follows swiftly if we close them.

Regardless of the nature of the excitatory stimulus, the behavior of the cortex cannot be dissociated from that of the thalamus because of the strong reciprocal nerve fiber connections linking the two structures. It was seen that neural activity originating in the cortex could spread to the thalamic nuclei and then be fed back to the cortex, setting up a continuing oscillation. Or, in the case of the well-known alpha rhythm, the thalamus was the source, only for the cortex to follow suit. It was because of the intimacy of the connections between thalamus and cortex that both Mircea Steriade and Rodolfo Llinás have insisted that one should think in terms of thalamo-cortical units, rather than of two independent structures. The activity of the gnostic units in the cerebral cortex, then, depends to a great extent on their thalamic connections.

But what is it that the thalamic neurons do? In the case of the specific thalamic nuclei, their main role is clear—to transfer information to the primary sensory receiving areas in the cortex. But the function of the nonspecific nuclei is less apparent though Rodolfo Llinás (Box 12.1) and his colleagues have pointed out that, at the local level, the coincidence of inputs from the specific and nonspecific nuclei could be important in bringing the pyramidal cells to the firing point.[25]

The role of the corticothalamic oscillatory activity is less obvious. One suggestion, originally made by Andersen and Andersson,[26] is that rhythmic EEG activity reinforces synaptic activity incurred in the acquisition of information,

Box 12.1 Rodolfo Llinás (1934–)

In an era of ever-increasing specialization, Rodolfo Llinás remains one of the last great all-round neurophysiologists, his work ranging from mathematical modeling to membrane conductances, single cell studies in animals and brain slices, intact human brain recordings, and much more. Llinás was born into a strongly medical family in Bogota, Colombia, where his father was a thoracic surgeon. After completing high school, Rodolfo entered medical school at 17, graduating in 1959. As a boy he had been interested in the brain, and this led him to choose neurosurgery as a specialty and the Massachussetts General Hospital for his training. After concluding that neurosurgeons tended to be too busy with clinical matters to explore brain function, Llinás switched to basic neuroscience instead, a decision facilitated by research he had been able to observe at MIT and Harvard. The next step was a move to Minneapolis where, in the department of Carlos Terzuolo, he devised and

Figure B12.1. Rodolfo Llinás. An enormously experienced neurophysiologist who has made original studies in many fields. (Author: Rodolfo Llinás, from Wikimedia Commons)

undertook neurophysiological studies of descending inputs to motoneurons. Having discovered that inhibitory inputs could involve synapses on dendrites as well as on the cell bodies, Llinás was encouraged to undertake graduate studies in Canberra with Eccles. There, working on the cerebellar cortex with Kazuo Sasaki, Llinás elucidated the circuitry of the connections to the prominent Purkinje cells. Returning to Minneapolis with a PhD, Llinás continued his study of the cerebellum before accepting an invitation to join Eccles in the latter's move to Chicago. Still with the cerebellum, Llinás noted that its neural structure was consistent across all the vertebrate species examined. Another major finding was that excitation of the Purkinje cells was mediated by propagated calcium spikes in the dendrites. He also investigated function in the inferior olive, the source of the cerebellar climbing fibers. Quite separate from the cerebellar work was research undertaken in summers spent at Wood's Hole, Massachussetts; there Llinás was able to work out the calcium kinetics in the presynaptic terminal of the squid giant axon.

Obliged to leave Chicago following Eccles' abrupt departure, Llinás took his research team to Iowa City and, after six years there, made his final academic move—to New York University School of Medicine as chair of physiology and medicine. Since that time (1976), Llinás has continued his extraordinarily varied successes, the latter including brain-slice studies of neural oscillations between thalamus and cortex, as well as pioneering investigations of the human brain with magnetoencephalography.

thereby consolidating memories. Although this suggestion seems rather unlikely— the pyramidal cell synapses involved in the memorizable events will be different from those made with the nonspecific thalamic fibers—the role of the alpha rhythm in cognition continues to attract support.[27] Another possible function of rhythmic cortical activity is that, since it may be associated with corticospinal tract discharges,[28] it provides a continuous descending control of brainstem and spinal cord activity.

The most widely held view at present, however, is that originally proposed by Christof von der Malsburg and by Wolf Singer and Charles Gray, namely that 40 Hz (gamma) oscillations provide functional links between different cortical areas engaged in the same task.[29] However, this "binding" hypothesis would be inconsistent with the time-chunk mechanism outlined earlier, since the chunks would be too brief for gamma rhythms to develop.

Very relevant to the interpretation of the gamma rhythm are the experimental results of Rodolfo Llinás and his colleagues in New York University. During electrical stimulation of mouse brain slices, they observed that simultaneous stimulation at the gamma frequency restricted excitation to a small focal area of cortex.[25]

Since a surge in gamma EEG activity can be seen in human subjects after sensory stimulation, as the same authors demonstrated using magnetoencephalography, it is possible that this, too, is inhibitory and serving to restrict the spread of excitation. Such a mechanism would provide a ready explanation for the inability to attend to more than one sensation (or thought) at a time.

Upon reflection, the need for continuing restraint on the cortex seems obvious—without it, spontaneous neural activity could give rise to all manner of spurious pains and illusions, most obvious in the patients with migraine and epilepsy. And even normal subjects may experience spontaneous pains and itching, flashing spots of light on closing the eyes, and momentary ringing in the ears. The same tendency for self-excitation can be observed outside the body for, when maintained in culture or grown from stem cells, mature neurons will readily excite themselves and their neighbors.[30,31] A sequence of IPSPs, as part of an EEG rhythm, would therefore be a means of providing the necessary dampening of the cortex, at the same time improving the "signal-to-noise" ratio in favor of purposeful electrical activity. The economy and effectiveness of the system would lie both in the long durations of the IPSPs and in the marked divergence of the basket cell axons, such that very few cells can inhibit hundreds of "burst" pyramidal neurons. Moreover, a single impulse from a burst neuron would be sufficient to excite a basket cell and bring the widespread inhibition into play.

The same argument could be made for the slower EEG rhythms, those observed during mental relaxation or deep sleep. In these situations the diminished spontaneous activity in the cortex could be dealt with by more prolonged IPSPs. As a consequence the time-chunks (generated by the layer V burst pyramids; see earlier discussion) would become too infrequent for consciousness to be maintained.

Finally, a role for the nonspecific thalamic nuclei in generating or supporting inhibitory cortical rhythms makes teleological sense. If the cortex evolved to receive excitation from the thalamus, that excitation could best be controlled by the same source.

Notes and References

1. See Chapter 9 for more information on Konorski.
2. Konorski J. *The integrative activity of the brain.* Chicago, IL: University of Chicago Press, 1967.
3. Sanghera MK, Rolls ET, Roper-Hall A. Visual responses of neurons in the dorsolateral amygdala of the alert monkey. *Experimental Neurology.* 1979; 63: 610–626.
4. Moscovitch M, Winocur G, Behrmann M. What is special about face recognition? Nineteen experiments on a person with visual object agnosia and dyslexia but normal face recognition. *Journal of Cognitive Neuroscience.* 1997; 9: 555–604.

5. Quiroga RQ, Reddy L, Kreiman G, Koch C, Fried I. Invariant visual representation by single neurons in the human brain. *Nature*. 2005; 435: 1102–1107.
6. Milner PM. A model for visual shape recognition. *Psychological Review*. 1974; 81(6): 521–535.
7. Mountcastle VB. Brain mechanisms for directed attention. *Journal of the Royal Society of Medicine*. 1978; 71; 14–28.
8. Demurget M, Reilly KT, Richard N, Szathmari A, Mottolese C, Sirigu A. Movement intention after parietal cortex stimulation in humans. *Science*. 2009; 324: 811–813 Andrew Schwartz (University of Pittsburgh) has also been a pioneer in this type of work.
9. See Chapter 6 in relation to the parietal lobe.
10. Eccles JC. A unitary hypothesis of mind-brain interaction in cerebral cortex. *Proceedings of the Royal Society B*. 1990; 240: 433–451.
11. McComas AJ. Containing the contents. In: *Consciousness: at the frontiers of neuroscience*. Jasper HH et al., eds. Advances in Neurology Vol. 77. Philadelphia, PA: Lippincott-Raven, 1998:135–148.
12. See Chapter 10 regarding the evidence for consciousness proceeding in "chunks" of time.
13. Thomson AM, Bannister AP. Interlaminar connections in the neocortex. *Cerebral Cortex*. 2003; 13: 5–14.
14. Cauller LJ, Kulics AT. The neural basis of the behaviorally relevant N1 component of the somatosensory-evoked potential in SI cortex of awake monkeys: evidence that backward cortical projections signal conscious touch sensation. *Experimental Brain Research*. 1991; 84: 607–619.
15. Stefanis C, Jasper H. Intracellular microelectrode studies of antidromic responses in cortical pyramidal tract neurons. *Journal of Neurophysiology*. 1964; 27(5): 828–854.
16. See Chapters 3 and 13.
17. Steriade M. Neocortical cell classes are flexible entities. *Nature Reviews Neuroscience*. 2004; 5: 121–134.
18. McComas A. *The artful chameleon: an exploration of migraine and medicine*. West Flamborough, ON: Alkat Neuroscience, 2006. See also McComas A, Upton A. Review. Therapeutic transcranial magnetic stimulation in migraine and its implications for a neuroinflammatory hypothesis. *Inflammopharmacology*. 2009; 17: 68–75.
19. Burns BD. Some properties of isolated cerebral cortex in the unanaesthetized cat. *Journal of Physiology*. 1951; 112: 156–175.
20. Buxhoeverden DP, Casanova MF. The minicolumn hypothesis in neuroscience. *Brain*. 2002; 125: 935–951.
21. Pape K. *The boy who could run but not walk: understanding neuroplasticity in the child's brain*. Toronto, ON: Barlow Books, 2016.
22. Meyers EM, Borzello M, Freiwald WA, Tsao D. Intelligent information loss: the coding of facial identity, head pose, and non-face information in the macaque face patch system. *Journal of Neuroscience*. 2015; 35: 7069–7081.
23. The Charles Bonnet syndrome was first described by the Swiss naturalist of that name in 1760. He had observed that his grandfather, losing his sight from cataracts, was subject to varied visual hallucinations. The condition affects rather less than 20% of

newly blind individuals, and the hallucinations, which may be of all manner of persons or objects, may last from seconds to hours. The hallucinations are an example of a "release" phenomenon. Hallucinations may also occur in healthy persons subjected to sensory deprivation, as described in the next chapter.

24. Hochberg LR, Bacher D, Jarosiewicz B, et al. Reach and grasp by people with tetraplegia using a neutrally controlled robotic arm. *Nature.* 2012; 485: 372–375.
25. Llinás R, Ribary U, Contreras D, Pedroarena C. The neuronal basis for consciousness. *Philosophical Transactions of the Royal Society of London B.* 1998; 353: 1841–1849.
26. Andersen P, Andersson SA. *The physiological basis of the alpha rhythm.* New York, NY: Appleton-Century-Crofts, 1968.
27. Sigala R, Haufe S, Roy D, Dinse HR, Ritter P. The role of alpha-rhythm states in perceptual learning: insights from experiments and computational models. *Frontiers in Computational Neuroscience.* 2014; 8: 36.
28. Whitlock DG, Arduini A, Moruzzi G. Microelectrode analysis of pyramidal system during transition from sleep to wakefulness. *Journal of Neurophysiology.* 1953; 16: 414–429.
29. Singer W. Synchronization of cortical activity and its putative role in information processing and learning. *Annual Review of Physiology.* 1993; 55: 349–374.
30. Corner MA, Crain SM. Patterns of spontaneous bioelectric activity during maturation in culture of fetal rodent medulla and spinal cord tissues. *Journal of Neurobiology.* 1972; 3(1): 25–45.
31. Amin H, Maccione A, Marinaro F, Zordan S, Nieus T, Berdondini L. Electrical responses of human iPS-derived neuronal networks characterized for 3-month culture with 4096 electrode arrays. *Frontiers in Neuroscience.* 2016 Mar 30; 10: 121. https://www.ncbi.nlm.nih.gov/pubmed/27065786.

13
Continuing the Synthesis

> We have seen the brain as an input-output signaling system. The signals entering it are not mental, nor are the executant signals which issue. But signaling which travels certain ways in the brain for instance through the great new nerve-nets seems to get, so to say, mental existence, though losing it again before even the penultimate exit-path. No microscopical, no physical or chemical means detect there anything radically other than in nerve-nets elsewhere. All is as elsewhere, except greater complexity . . . the spectacle drowns out any naïve notion that the activity of a single cell by itself can ever amount to a mental experience. For that, we have to seek rather some attribute of the organization itself.
> Charles Scott Sherrington, 1933[1]

Having considered cortical columns and gnostic units—the "building blocks" of cortical function—in some detail, it is time to step back and consider the brain as a whole. There is also more history to deal with, notably the work by experimental psychologists on reflex behavior. The "hard problem," too, deserves another examination, especially in the light of back-projections to the primary sensory receiving areas in the cortex. Finally, there are the intriguing observations on patients with "split" brains that cannot be ignored in any scientific discussion of the whereabouts of consciousness.

The first task is to tackle, once and for all, the idea that consciousness has an existence that is at least partly independent of brain activity ("dualism").

Dualism and Monism

In a human being the sense of "self" is so strong that it seems to be beyond the normal workings of the brain. It was this attitude that led Descartes to postulate a psychic component that interacted with the brain through the pineal gland. Though the latter concept may seem ridiculous now, more than one distinguished neuroscientist has been led to propose a modified dualism. In modern terms, this has meant invoking the existence of some factor beyond impulse-mediated activity of nerve cells. Thus Eccles conceived of quantum mechanics operating on molecules at the level of the synapse.[2] Quantum mechanics was also invoked by Roger Sperry, but only as an analogy for what might be happening in the brain to produce consciousness. In his view consciousness was generated by brain activity but then achieved an independence. Benjamin Libet was more specific, since he conceived of neurons generating some kind of influence surrounding the brain in his conscious mental field (CMF) theory. In his words, the CMF would be "the entity in which unified subjective experience is present." In addition, the CMF would have the ability to influence nerve cells. Although this emergent field would, by its nature, not be capable of examination by physical methods, Libet was nevertheless able to suggest a means of testing his hypothesis. The experiment would involve seeking evidence of mental activity in slabs of cortex that had been surgically or otherwise functionally isolated from the remainder of the brain.[3]

The alternative to the emergence of some new entity from neural activity is that the impulse activity itself constitutes consciousness. It was a conclusion of this nature that Thomas Huxley espoused in the famous lecture, cited earlier, during which he drew an analogy between the working of the brain and a steam locomotive:

> The consciousness of brutes would appear to be related to the mechanism of their body simply as a collateral product of its working, and to be as completely without any power of modifying that working as the steam-whistle which accompanies the work of a locomotive engine is without influence upon its machinery.[4]

In one sense Huxley was surely right—consciousness is a product of the brain machine—but in another sense his analogy is incomplete. Even at a very basic level, that of pain and pleasure, consciousness goes beyond mere reflex activity by playing an important role in shaping behavior, and it does so by modifying synapses. Neural circuits that are associated with pleasure are likely strengthened by repeated use; circuits resulting in pain will also become better established, again by the strong emotive component. Consciousness, in short, has been a hugely important evolutionary tool.

Aims and Limitations of a Synthesis

At this stage there are two important points to be made. The first is that, even though the continuing synthesis is necessarily incomplete and is certain to be incorrect in part, the fundamental observations on which the synthesis is based must be included in any alternative scheme. This limitation includes the conjectures concerning gnostic units and cortical columns in the previous chapter, even though the speculations were, for the most part, based on firm observations.

The second point is that the "hard" question, how nerve impulses or other electrical potentials give rise to thoughts and to the qualitative aspects ("qualia") of sensations, may prove to be insoluble—at present, there is simply no experiment conceivable that could provide an answer. Though, as Patricia Churchland has emphasized, one cannot rule out a possible breakthrough in the future,[5] there is presently a strong case for removing the issue as a question altogether. Either the cognitive processes of humans are unsuited to solving the problem, as Colin McGinn would have it,[6] or else the nerve impulse-qualia relationship is a fundamental property of certain neural networks and is incapable of being broken down. The situation should not give rise to total despair, however, for although the core of the qualia problem may be insoluble, some peripheral aspects *do* lend themselves to analysis, as described later.

The next proposition, one linked to the last, is also a major one—having developed the ability to acquire information about itself and its surroundings, any group of neurons will exhibit "consciousness." In its most rudimentary form this consciousness is simply an organism's awareness of itself and an ability to feel pain and pleasure; in the highest form it is the writing of a symphony or the formulation of general relativity theory. While we can make a reasonable guess as to the type of consciousness that a nonhuman primate or a dog may possess, we cannot conceive the nature of consciousness in a bat—as the philosopher Thomas Nagel famously pointed out[7]—or, for that matter, in an octopus or an insect. Yet, judged by their respective behaviors, they must all have some degree of consciousness.

Inseparable from the apparent indivisibility of consciousness and impulse activity is the concept of "free will."

Free Will

We feel that we have control over our thoughts and actions. The belief is a strong one. "I can continue word processing now or I can stop. If I stop I can make myself a meal or go for a walk. If I decide to walk, I can take my usual route through the nearby park or go around the block instead. But perhaps I should telephone my wife first, to let her know what I am about to do," and so on.

But there have long been doubters about free will, one of whom was Baruch Spinoza (1632–1677). Spinoza, born in the Dutch Republic, was a noted philosopher who included Descartes' treatises in his studies. Unlike Descartes, however, he found no place for dualism and saw the situation very differently: "Men are conscious of their desire and unaware of the causes by which (their desires) are determined."[8] Had he been alive today, Spinoza would not have been surprised by the result of Benjamin Libet's famous experiment, an outcome that, more than any other observation, appeared to destroy the concept of free will altogether. Libet, it will be recalled, was able to record a potential from a subject's brain that preceded the conscious intent to make a movement. Although a number of criticisms were raised afterward, the basic conclusion stood. Thus, rather than the mind bringing the motor apparatus of the brain into play, it was the other way round—the neural activity created the sense of "will." Libet's experiment has provided the best evidence against any form of dualism.

But why is the feeling of control over our thoughts such a powerful one? The answer lies partly in the fact that the brain is making "decisions" on the basis of past and present factors, some of which reach consciousness. If the brain stops my word processing now, for instance, it is because there are hypoglycemic messages reaching it, together with impulses in somatosensory neurons conveying information about body posture and other impulses encoding the memory of a prepared meal waiting in the dining room. Each type of message enters my consciousness, giving me the illusion that, because I am tired and hungry and know that there is a meal set out for me, I have decided to stop word processing and to eat instead.

Similarly, if one considers a more remote problem—the planning of a holiday, for example—there is a host of neural assemblies (Konorski's "gnostic units") in the brain coding for different places at various times of the year, possible traveling companions, snorkel and ski equipment, and so on. All of these assemblies have strong or weak links to each other, and the effectiveness of these connections will determine which assemblies will bind together so as to reach consciousness as an (illusory) decision.

To reiterate, there can never be "free will" because the brain will always be influenced by past and present factors, as represented by specific neural groupings. From the time that an infant first opens his or her eyes, if not before, there is no such thing as an open mind.

Human Behavior as Reflexive in Nature?

If free will is an illusion, should behaviors in humans, and those in other animal species, be regarded as purely reflexive? One of the first neuroscientists to reach this conclusion was Ivan Sechenov (1829–1905), whose work in identifying inhibitory processes has already been mentioned.[9] That human behavior originated in reflexes was the subject of his influential essay published in 1863.

This idea, that, despite appearances to the contrary, human activity was explicable on the basis of reflexes, found a strong proponent in another Russian, Ivan Pavlov (1849–1936; Box 13.1). Like Sechenov, Pavlov was born in a small village

> ### Box 13.1 Ivan Pavlov (1849–1936)
>
> Ivan Petrovich Pavlov had at least three claims to fame—as a pioneer of classical conditioning, as a survivor of the Stalin purges, and as the first Russian scientist to win a Nobel Prize. Throughout his teens, however, it seemed very likely that Pavlov would follow his father into the Russian Orthodox priesthood and he had, in fact, enrolled in the seminary in his hometown of Ryazan. One of those who, through his writings, influenced Pavlov to abandon the church was the "father of Russian physiology," Ivan Sechenov. Inspired by science, Pavlov left Ryazan for St. Petersburg, where his university studies included natural sciences. A fourth-year project on the innervation of the pancreas was well received and led to a junior appointment at the Academy of Medical Surgery, and then to an assistantship in the physiology department at the Veterinary Institute. During this phase of his career Pavlov investigated the circulatory system and also became intrigued by reflex activity and by the trophic (sustaining) action of nerves on tissues. After winning a gold medal for research, Pavlov decided to further his career as a physiologist by spending
>
>
>
> **Figure B13.1.** A young Ivan Petrovich Pavlov in a thoughtful mood. Even though he was critical of the postrevolutionary government, Pavlov's fame and international standing helped him survive the Stalinist purges of the 1930s.

two years in Germany, initially studying under Carl Ludwig in Leipzig. However, it was his experience with Heidenhain in Breslau that introduced him to the use of exteriorized pouches of stomach or intestine to investigate secretion of digestive juices. Pavlov was able to improve the surgical technique so that nerves to the pouches were left intact. Further, he developed a technique for collecting secretions from externalized salivary glands. It was while working with the latter preparation in dogs that Pavlov found that the animals, having received a warning cue (ringing a bell or sounding a buzzer), would salivate in anticipation of being fed. This was the first description of classical conditioning, and the relevance of the principle to many human situations was soon obvious to psychologists.

At the time of his successes in the laboratory, Pavlov held the chair of physiology at the Medical Military Academy in St. Petersburg as well as that at the Imperial Institute of Experimental Medicine. His numerous students included Jerzy Konorski (of "gnostic units") and several others who would achieve fame. Probably because of his Nobel Prize, awarded in 1904, and his considerable international reputation, Pavlov survived the widespread Stalin purges in the 1930s—this despite his public criticisms of the Soviet government.

to the east of Moscow. And like the older man, would move to St. Petersburg to pursue his career. During the years between 1876 and 1891 the two physiologists were in different institutes but would nevertheless meet to discuss their work. Pavlov had high regard for Sechenov, referring to him as the "father" of Russian physiology and he would certainly have been familiar with Sechenov's earlier essay on "Reflexes of the Human Brain." Where Pavlov surpassed Sechenov was in his experiments, especially in those showing that reflex behavior could be learned by the nervous system. The best-known example of such a conditioned reflex was the dog that, having been trained to receive food after the ringing of a bell, would salivate in response to the sound of the bell alone.

The work was taken a step further by two young Poles who showed that reflexes could also be formed without the elaborate conditioning employed by Pavlov. Rather than using the secretion of saliva as an end-point, they chose movement of a leg instead. One of the young men was Jerzy Konorski who, in later years, developed the concept of "gnostic units" discussed in the previous chapter and again here.

Although Pavlov went on to win widespread fame, together with the 1904 Nobel Prize in Physiology or Medicine, the work of Sechenov, Konorski, and other Russian psychologists attracted little attention in the West until much later.

Instead it was the turn of two Americans to develop the idea that human behavior was mediated by acquired reflexes. The first was John (JB) Watson (1878–1958).

Watson, the son of an alcoholic father, had an undistinguished record at school. He was nevertheless admitted to a local university and then went on to doctoral studies in psychology at the University of Chicago. Aware of the studies of Pavlov, Watson became a strong advocate of "behaviorism" in the study of psychology; it was a philosophy that regarded the conscious reporting of introspections as an unreliable experimental approach and that argued instead for the examination of behaviors by external observers.[10]

The second American, a younger man, had an even greater influence for the understanding of consciousness and the rejection of free will and, like several other major contributors to the field of consciousness, had done so while a member of the Harvard faculty. B.F. Skinner (Box 13.2) grew up in the Pennsylvania coal-mining county of Susquehanna. After studying literature for a bachelor of arts degree, he went to Harvard where, having read JB Watson's *Behaviorism*, he decided to pursue experimental psychology for a PhD. While still a graduate student he invented an apparatus for measuring animal behavior—the "Skinner Box" (Figure 13.1). Like Watson before him, Skinner denied the existence of free will, positing instead that behavior, rather than being consciously directed, resulted from the automatic response of the nervous system to environmental influences. In their simplest forms, behaviors could form in the way that Pavlov had described in his work on conditioned reflexes. Alternatively, behaviors would develop if activities came to be associated with, and reinforced by, situations resulting in positive (pleasure) or negative (fear, pain) feelings; this process was termed "operant" or "instrumental" conditioning.[11]

Though he was probably unaware of it, operant conditioning had been described earlier by Jerzy Konorski and Stefan Miller in Poland in their bold student experiments on a single dog. Skinner recognized, however, that in human subjects most behaviors are much more complex than the ones he described and he suggested the involvement of "chains" of conditioned associations as an explanation. One of the most convincing results of Skinner's approach was his success in creating superstitious behaviors in animals. He was able to achieve this by rewarding the animal every time it made a particular "spontaneous" movement; as an example, one of his pigeons would regularly rotate three times in the expectation that food would suddenly appear.

Modern neuroscience has helped to firm up the concept of conditioning by revealing the importance of the neurotransmitter that Arvid Carlsson discovered—dopamine—in the acquisition of pleasurable traits and addictions. The basal ganglia and limbic system are also involved, the former to do with movement in the behavior and the latter providing an emotive component.

Box 13.2 B. F. Skinner (1904–1990)

Burrhus Frederic Skinner did not set out to become the most influential psychologist of his time. He was born in Susquehanna, Pennsylvania, where his father was a lawyer. After completing high school, Skinner attended Hamilton College in New York State with the intention of making his career as an author. Having gained a bachelor's degree in English literature, Skinner—still aiming to write—moved to Boston for graduate studies at Harvard. While there, he was persuaded by Fred Keller, a fellow graduate student, to explore behavior as a science, and the two of them devised a number of experimental tools for this purpose, the most important being the "Skinner Box." There were two such devices, one for rats and another for pigeons. In both instances a particular activity on the part of the captive animal—pushing a lever for a rat and pecking at a target for a pigeon—was rewarded with food but, in most experiments, only if the animal had performed the action in response to a signal. This type of behavior became known as operant conditioning.

Figure B13.2. B.F. Skinner at Harvard, c. 1950. (Wikimedia Commons CC 3.0; author Silly rabbit)

> To complement the observations of the experimenter, the actions of the animal were recorded automatically by an ink-writer. Despite the almost instant success of his inventions, it was only after failing as a writer that Skinner decided to pursue experimental psychology as a career. In part, his decision was influenced by his introduction to J. B. Watson's *Behaviorism*. Having obtained his doctorate in psychology and spending five years in research, Skinner left Harvard to teach at the University of Minneapolis and then at Indiana University. In 1948 Skinner returned to Harvard as a tenured professor and remained there until his retirement in 1974.
>
> A highly successful academic and one well-known to the public, Skinner was a controversial figure because of his conviction that human behavior, like that of the animals he studied, consisted of actions that had been shaped by pleasure or dislike on exposure to the same stimuli on previous occasions ("reinforcement"). Humans, too, operated on a reward system, and free will, contrary to popular belief, was an illusion.
>
> Though he had not succeeded in his early aim to become a novelist, Skinner nevertheless became a prolific author, expressing his ideas to the general public and also attempting to influence education in schools. Regarded as the greatest psychologist of his generation, Skinner was the recipient of numerous awards during his lifetime.

Returning to the earlier question—whether all human behavior should be regarded as reflexive—the answer is surely dependent on the behavior concerned. If the same stimulus brings about an unvarying response in a conscious individual then, yes, such behavior is clearly reflexive. Suitable examples would include such basic activities as coughing, sneezing, yawning, and scratching—even though these can be suppressed by an effort associated with consciousness. Reaching for a glass of water when thirsty or closing the eyes when concentrating would be examples of less stereotyped but nevertheless reflexive actions. Putting on the same hat and inspecting the garden before setting off on a daily walk would represent yet more complex instances. But there are so many situations, both in humans and in higher animals, where the large number of gnostic units operating in the cortex ensures that the outcome of a stimulus cannot be predicted by an external observer. Apart from this, it is very likely that some daily thoughts and actions occur by chance—the outcome of spontaneous firing of the neurons comprising the gnostic units. It is this multiplicity of possibilities that fosters the illusion of free will. Therefore, in the sense that a reflex describes an unvarying response to the same stimulus, much—perhaps most—of human behavior is not reflexive, even though there has been no conscious intervention

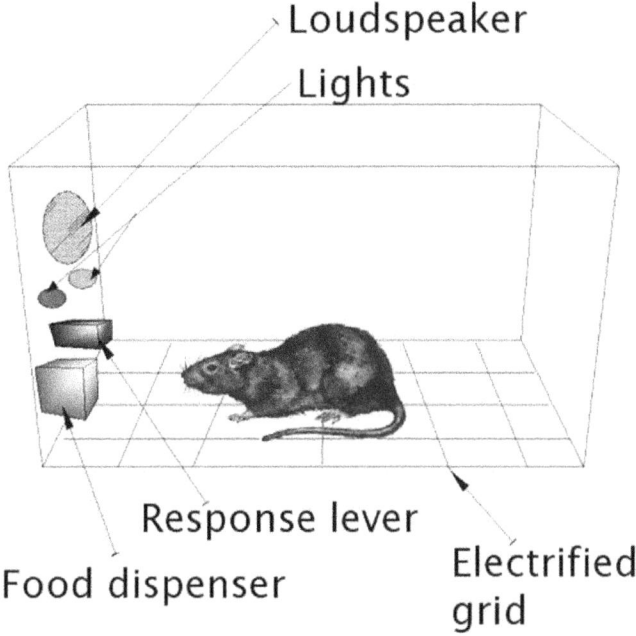

Figure 13.1. The Skinner Box. The box can be used in several ways. In the simplest the rat learns that pressing a lever releases a food pellet from the dispenser (positive reinforcement). Alternatively a rat can discover that pressing a lever stops a painful electric current to its feet (negative reinforcement). More complex experiments involve warning lights or sounds. (Wikimedia Commons. Author: Andreas 1, adapted from image: Boite skinner.jpg https://commons.wikimedia.org/wiki/File:Skinner_box_scheme_01.png)

in the brain's decision-making. Though not reflexive, the behavior is still independent of "will," however.

A moment's reflection should help to dispel any resistance to the idea of the brain performing most of its work, even high-level work, unconsciously. We are aware of those occasions when a forgotten name suddenly springs to mind, or when the answer in a crossword puzzle starts to be entered before the word reaches consciousness. Often the brain will present its most creative work when least expected. The ring structure of benzene allegedly came to Auguste Kekulé while dozing by the fire, while Jules-Henri Poincaré envisaged a new mathematical function as he was about to step into a cab. In neuroscience, the design of an experiment for proving the existence of a neurotransmitter awoke Otto Loewi from a dream. Dreams were also responsible for Tartini's famous violin sonata, "The Devil's Trill," and for musical compositions by Schumann, Stravinsky, and

Wagner.[12] And, at a more mundane level, the automatic machinery of the brain is capable of driving a car through a busy city or playing a musical instrument while the thoughts of the motorist and musician are elsewhere. Indeed, it was in relation to the latter that Charles Darwin made a relevant comment concerning what he and his contemporaries termed "double consciousness." Writing of his future wife, Emma Wedgewood—an accomplished pianist who had studied with Chopin—Darwin noted: "'Emma W says that when she is playing by memory she does not think at all, whether she can or cannot play the piece, she plays better than when she tries."[13]

In addition to the removal of free will as a scientific problem, there is another gain and this concerns attention. In bringing the content of a gnostic unit to consciousness, the automatic working of the brain has, at the same time, created the impression that it is the subject of willed attention. Inevitably there must be some particular feature of the gnostic unit involved—perhaps a high-frequency burst of impulses—but there is no conscious direction of attention.

Back to the Hard Problem

It is accepted among neuroscientists that, in relation to consciousness, the outstanding difficulty is not so much the neural circuitry involved but, rather, the nature of the process whereby electrical activity in neurons is transformed into a sensation—the redness of a rose, say. While it likely that this is a problem not amenable to solution, it is nevertheless instructive to approach it with the following considerations.

The Sensory Receptors Specify the Cortical Modality

The first point is that the capacity to produce the sensation—the color, the sound, the touch—is something that is ultimately endowed upon the cortex by the receptors involved, be they cones in the retina, hair cells in the cochlea, or mechanoreceptors in the skin. This is not necessarily as obvious as it might at first seem. One could imagine, for example, a developing brain in which certain cortical areas were already programmed to create a given sensation and that the role of the receptor pathway in the fetus and neonate was to organize the committed cortical neurons so as to give meaningful responses to given stimuli. In tissue culture and fetal animal experiments, however, it has been shown that this is not the case—it is the developing thalamocortical axons that, on reaching their target, decide the nature of the sensation to be generated by the cortex.[14] To quote Colin Blakemore, the fetal cortex is a "tabula rasa." In keeping with this conclusion are the many examples of neuroplasticity considered previously, in which deafferented cortex is taken over by other sensory inputs.

The Back-Projections Produce the Greater Part of a Sensory Experience

In Chapter 10 evidence was presented for functional back-connections between the "higher" cortical visual cortex (V3, V4, V5, inferior temporal lobe) and the primary visual receiving areas (V1, V2). The suggestion was made that the back-projections were responsible for eliciting color, detail, and possibly movement in the visual perception, while the inferior temporal lobe, through experience, recognized shapes.

The tripartite process described for vision applies equally well to other sensory modalities. Regarding somatic sensation, microelectrode studies on the primary somatosensory cortex (SI) strongly suggest the presence of a back-projection from higher areas (SII, Brodmann areas 5,7).[15] The primary area might then be responsible for sensing "touch," say, while a higher area, incorporating the body schema, would signal "where." Similarly, in the auditory system, the primary cortical receiving area would be responsible for the "quality" of sounds. As an example, suppose we meet someone and exchange greetings—"Good morning" . . . "Nice day" . . . "How are you?" The higher auditory area in the temporal lobe will have gnostic units specific for each of these common phrases and so recognizes them. But how the phrases actually sound—whether they are spoken loudly or softly, slowly or quickly, by a child or by an adult—is the outcome of the neural activity in the primary auditory area.

The idea of back-projections being responsible for a conscious sensation is such a powerful one that it is helpful to consider an analogy from outside the brain.

The Enigma Machine

As a prelude to the analogy, imagine using a microelectrode to identify the column(s) of neurons in the human temporal lobe that discharges whenever a particular face is seen. Then imagine, with very fine dissection, removing this column from the brain. Clearly, that face can no longer be recognized by the individual since its "grandmother cells" are now missing. But it is also highly unlikely that the isolated column, now resting in a culture medium, would be able to recall the face when stimulated electrically. The gnostic unit—the cortical column—can only be effective through its connections in the intact brain.

Enigma was a cryptographic device used by the German military in World War II.[16] (It was the breaking of the Enigma codes that was the triumph of Alan Turing and his associates during the war; see Chapter 14.) The ingeniously conceived Enigma machine was able to disguise a message by replacing each letter of the alphabet with one that had been determined by a series of interactive rotors (Figure 13.2). As an example, the letter G on the typewriter keyboard was first connected by wire (through a plugboard, not shown in the figure) to the letter W on the static rotor (wheel) on the right. From then on the message was passed on from one rotor to the next by electrical contacts rather than wires. Further, on each wheel, the inner and outer rings of contacts could be set independently. In the figure, W on the static rotor has become A on the outer ring of the adjacent rotor and then Z on the inner ring of the same rotor. On the next rotor, the Z becomes B and then Q, while on the final rotor the Q has been replaced by S and then T. So, after passing through four rotors, the G on the typewriter keyboard has become T on the final rotor—T is now the code for G. However, this is merely the halfway stage in the coding procedure. In the Enigma machine the letter T would now become P on a "reflector" and begin a journey back through the same rotors, being scrambled as it passed from one to the next. Ultimately it would reach a "lightboard" and might now have become X. Thus G on the typewriter became T on the final rotor and then X on the lightboard. (As if that sequence of events did not pose

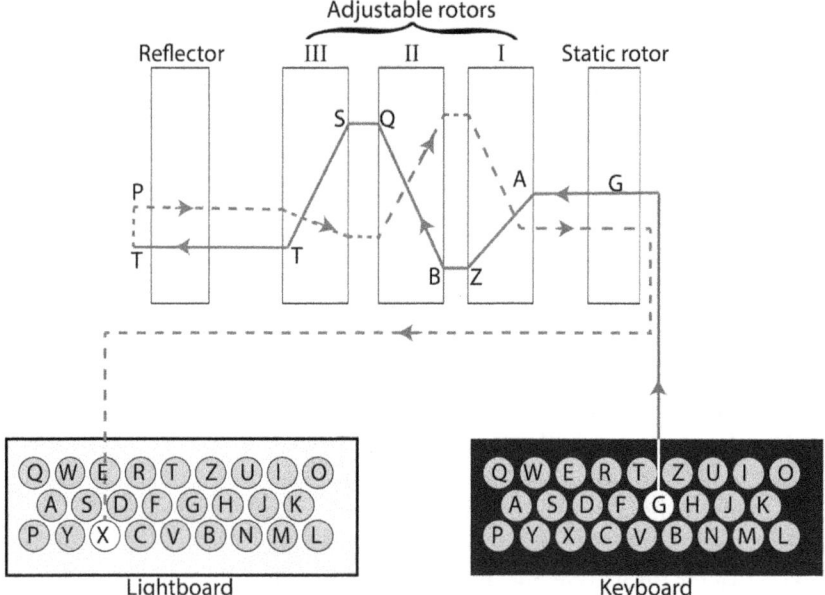

Figure 13.2. The principle of the Enigma machine.

sufficient difficulty for the British cryptographers, the Germans would turn the rotors to new positions each day.)

Now consider a sensory pathway, or "analyzer" as Konorski would refer to it. Replacing the typewriter keyboard, but acting like the keyboard, is a sensory receptor that converts the energy of the stimulus into an electrical signal, in this case one or more nerve impulses. The impulse message is then fed through a series of neurons in the brain, each group of neurons categorizing the incoming information and, like the Enigma rotors, narrowing the possibilities. Finally the impulse message arrives at the group of neurons that is the appropriate gnostic unit—corresponding to the final rotor of Enigma. The original stimulus is now coded as a gnostic unit.

Suppose now that, like the reflector in Enigma, there is activation of the gnostic unit from another group of neurons, say the nonspecific thalamic nuclei or perhaps another type of gnostic unit. Just like Enigma, the gnostic unit will pass its information back toward its peripheral source. Unlike the normal operation of Enigma, however, the path back will lead to the original starting point. (Had Enigma used the same path it would have resulted in the letter G on the Enigma keyboard showing up as G on the lightboard.) We could, if we wished, make the electromechanical analogue an even better fit by imagining a second output from the typewriter keyboard, this directly to the lightboard. A bulb on the lightboard would then be activated not only by the returning signal from the most distant rotor but by the depression of the appropriate typewriter key (Figure 13.3).

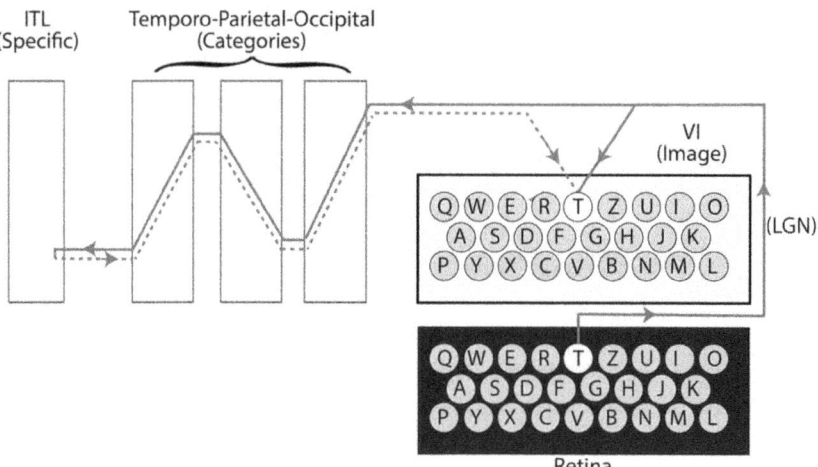

Figure 13.3. Modification of Enigma machine to resemble visual processing. *ITL*, inferotemporal lobe; *LGN*, lateral geniculate nucleus.

Although we have now so modified Enigma that it is no longer a secret coding device, the final version simulates a number of features of a sensory pathway, that is:

> Like a receptor it transforms environmental energy (depressing a key) into an electrical signal.
> The signal passes through several stages (series of rotors).
> The signal can be stored (final rotor).
> The signal can be recalled (reflector).
> Recall reactivates the original response (lightboard).

Advantages of Back-Projection

It should be immediately obvious that, in the use of back-projections, the nervous system is simplifying its sensory task. Rather than having to convert environmental energy into nerve impulses at the receptors and then, by complex neural transactions, into a sensation in "higher" cortical areas, only a modified initial step is necessary. In such a scheme the main function of the subcortical afferent pathways is to transmit the impulse activity from the receptors to the cortex, rather than to modify it significantly. Hence the concentric center-on and center-off fields of the retinal ganglion cells resemble the fields of cells in the lateral geniculate body and in the V1 cortical blobs. Similarly the responses in peripheral nerve fibers to touch and pressure on the skin are similar to those in neurons of the primary somatosensory cortex. It is as if the back-projections were rekindling activity in the receptors (skin) or receptor-complex (retina) themselves. Put another way, a nerve cell will only "know" red, irrespective of whether it is being stimulated by electromagnetic radiation of a certain wavelength or by impulses from another part of the brain.

Back-Projections in Dreams and Imagination

The double role of the primary cortical receiving areas in generating sensation (transmission and re-excitation) is also supported by clinical observation, as was recognized many years ago. Thus, in an extraordinarily perceptive article, only two pages long and appearing in an obscure journal in 1893, T. M. Balliet wrote:

> we recall things seen with the very cells with which we see ... we imagine color with the same cells with which we see and recall color .., but a person whose

blindness is due to an injury to the optic brain center, does not simply not see, but such a person cannot recall anything he has ever seen, and cannot imagine anything he has ever seen.[17]

Balliet applied his reasoning not only to vision but to all the senses. In relation to hearing, he conjectured: "Beethoven composed music after he was deaf. His deafness must have been ear-deafness. If his auditory centers had been injured, his imagination of tone would have gone."

One might add that the converse is also true. If we wish to imagine something we close our eyes so that the cells in the visual receiving areas (V1, V2) can be employed for the imagined scene rather than for processing information from the external world.

At the time that Balliet was writing, the posterior occipital lobe had been identified as the visual area of the brain, but in a patient there was only evidence from the clinical examination and from neuropathology to indicate where a lesion might be. Nowadays, however, the location can be determined with exquisite accuracy by magnetic resonance imaging and the latter has, in fact, been employed to demonstrate that ischemic infarction of "both deep occipital lobes" is associated not only with lost vision, as would be expected, but also with an inability to dream.[18]

How Did Sensory Consciousness Evolve?

While the preceding discussion deals with the necessary conditions in the cortex for the redness of a rose to be perceived, it does not touch the essence of the hard problem, namely the nature of the transformation of the electrical activity in the primary visual area (V1) into "redness." One attempt at a solution would be to tackle the issue with an evolutionary approach. Thus, in certain early single-celled life forms, photosensitive pigments would have appeared in the cytoplasm or in the membranes. As more complex organisms evolved, special cells housed the photopigments. Initially the pigments did no more than signal to the organism that the environment had switched from dark to light, or vice versa. In a species with a primitive awareness (see later discussion), however, a light signal would eventually come to be associated with a reward—the ingestion of food, for example, or a sexual mating. In this way the primordial consciousness would gain a new dimension, one capable of distinguishing light from dark. Suppose now that a second type of photosensitive pigment and receptor cell evolved, one that responded to a wavelength spectrum different to that of the first photoreceptors. Activation of this second type of receptor would not be expected to yield the same sensation as that elicited by the earlier type, as a different population of cortical cells would be involved. Thus if the earliest photoreceptor had come to evoke the sensation of whiteness, activation of the later-evolved receptor would have to produce a different consciousness—say that of green or blue or... the redness of a rose.

In summary, although the various types of speculation do not get at the heart of the hard problem, they do at least bring it more clearly into view.

Which Areas of the Brain Contribute to Consciousness?

One of science's most tantalizing questions— the "where" of consciousness— has exercised laboratory workers and philosophers alike from the time of René Descartes and Spinoza to the present era.

Lashley and "Mass Action"

One school of thought supposes that all cortical areas make equal contributions to consciousness, and it is likely that, to some extent, this attitude arose from an extrapolation of the work of Karl Lashley (Figure 3.3). Lashley, like Watson and Skinner, was an influential Harvard psychologist with considerable expertise in animal experimentation. The main object of Lashley's studies was to locate the site for memory (the "engram"). His experiments involved making selective cortical lesions in rats that had been trained to travel through a maze. It appeared to him that, at least in this species of mammal, no one area contained the engram since the rats could find their way regardless of the site of the cortical destruction—a conclusion subsequently overturned by the discovery of "place" cells in the rat hippocampus. His ablation experiments also led Lashley to develop a theory of "mass action," a theory that postulated that success in learning depended on the amount of cortex available rather than on the activity of a specific locus. Although Lashley had not investigated consciousness itself, it was perhaps understandable that his conclusions about the equivalence of cortical regions led some to suppose that they applied to consciousness as well as to learning.

Variability of Cortical Contributions

The position taken here is that the various cortical areas differ greatly in their contributions to consciousness. At one extreme are the frontal lobes, including the most anteriorly situated "prefrontal" areas. While accepting the importance of the precentral cortex for the execution of movement, and of the prefrontal areas for making judgments that influence complex behaviors, the fact remains that human subjects remain fully conscious after large parts of the frontal lobes have been injured, surgically removed, or disconnected from the remainder of the brain. In terms of gnostic units, it would appear that, despite the huge numbers of cortical columns available, relatively few gnostic units had been formed during the preceding lives of the patients. In contrast, the effects of damage to more posterior parts of the brain are severe. If the damage involves the territory of a sensory analyzer there may be blindness, deafness, or an inability to comprehend speech; if the lesions are more restricted, the loss of function may be more subtle—for example, an inability to recognize faces or objects or to name them. Yet, no matter how incapacitating the plight of the patient, consciousness is still present.

Effects of Near-Total Sensory Deprivation

As a side issue, one might inquire as to the effects on consciousness of removing not one, but all, external senses. The first such sensory deprivation experiments appear to have been carried out in Donald Hebb's psychology department at McGill University in Montreal in the 1950s. Paid volunteers were isolated in small rooms, lying on narrow couches with pillows covering both ears and with frosted goggles over the eyes to remove patterned visual stimulation.[19] Later experiments, by Lillie in California, removed skin sensation as well, this being achieved by having the subjects floating in a bath of warmed Epsom salt solution.[20] The results of the sensory deprivation in both series of experiments were not a loss of consciousness but an unexpected development—the onset of visual hallucinations.[21]

Loss of the Body "Schema"

Even though functioning sensory areas are not required for consciousness, there is one area in each hemisphere that might be thought essential, and this is the region in the parietal cortex that contains the gnostic units for the "body schema" of Henry Head. Thus a lesion of this area on the right side typically results in a lack of awareness of the left side of the body and its surroundings. Though the effects of damage to the corresponding region on the left side of the brain are less severe, this has been attributed to the right side of the body being represented in both hemispheres. Pertinent observations bearing on this matter have also come from the study of a patient with an especially severe form of paralytic migraine. While it is known that some migraineurs may experience distortions in the perceptions of their bodies, in this patient there were also occasions when some or all of her body "disappeared" (e.g. "I'm only a head!" and "The whole of me is gone!") mere seconds prior to the loss of consciousness.[22] As important as this particular parietal area is for full consciousness, it clearly is not sufficient in itself, for the migraineur was still aware of her body's absence and could state the situation clearly. On the other hand, when the entire cortex is suddenly disabled, as by the "spike-and-wave" EEG discharges characteristic of "absence seizures," consciousness is lost.[23]

Ability to Experience Pleasure and Pain

The approach taken here in the search for the locus of consciousness is to focus on pleasure and pain—basic and contrasting feelings essential for learning and survival. In the case of pain, its importance is evident from the lives of those human subjects unfortunate enough to have been born with an inability to experience it, due either to a sodium channel abnormality or an absence of unmyelinated peripheral nerve fibers. Affected individuals stumble through life burning their hands, breaking bones, and generally injuring themselves—all because they never have the warnings that come from painful prior experiences. Pleasure is a normal accompaniment of, and hence a drive for, the essential activities of mating and feeding. Since the experience of pain and pleasure requires a conscious nervous system, one can argue that consciousness depends on the presence of the limbic system, since it is the latter that contains the neural structures that, in humans and other mammals, are most active in responding to the requisite stimuli. Especially important within the limbic system are the anterior cingulate gyrus for pain and the nucleus accumbens for pleasure.

The argument here can be extended by proposing that "primitive" creatures that have neural assemblies responsive to stimuli evoking pain and pleasure in more highly developed species are also likely to have consciousness. Extending the argument, it is even conceivable that an insect experiences a primitive consciousness as it responds to the attraction of a powerful pheromone.

In humans, the emphasis on pain-pleasure mechanisms underlying consciousness is not necessarily inconsistent with Antonio Damasio's proposal that, in the lower neuraxis, in the dorsal regions of the midbrain and pons, exist neural assemblies capable of generating the "primordial" feelings of a "protoself."[24] Although it is necessarily circumstantial, one line of evidence pointing to the brainstem as a source of primitive consciousness is the carefully documented behavior in anencephalic infants suggestive of pleasure and pain.[25,26] Finally, looking beyond Damasio, there is the almost forgotten figure of Henry Head with his concept of "epicritic" and "protopathic" nervous systems. It was Head who, following years of studying somatic and visceral sensation in himself and his patients, identified the thalamus, as the head-station of the protopathic nervous system, as responsible for a basic, poorly defined consciousness.[27]

In summary, then, it is likely that there is a basic consciousness originating in the brain stem and limbic system, one that includes feelings of pain and pleasure as well as a vague awareness of self. On this primordial consciousness the various cortically mediated consciousnesses are superimposed—the special senses, thoughts and language, the body schema, autobiographical memory, reasoning and planning. Just as there is a hierarchy in the structure of the nervous system, so is there a hierarchy of consciousness. Figure 13.4, which is an extension of an earlier illustration,[28] summarizes the situation. In this figure no particular function has been assigned to the prefrontal cortex. From lesion studies we know that the prefrontal areas influence mood and judgment and there is neurophysiological evidence that some "working" memory is carried out there, possibly assisting function in more posterior cortical regions—but that is all, for the moment at least.

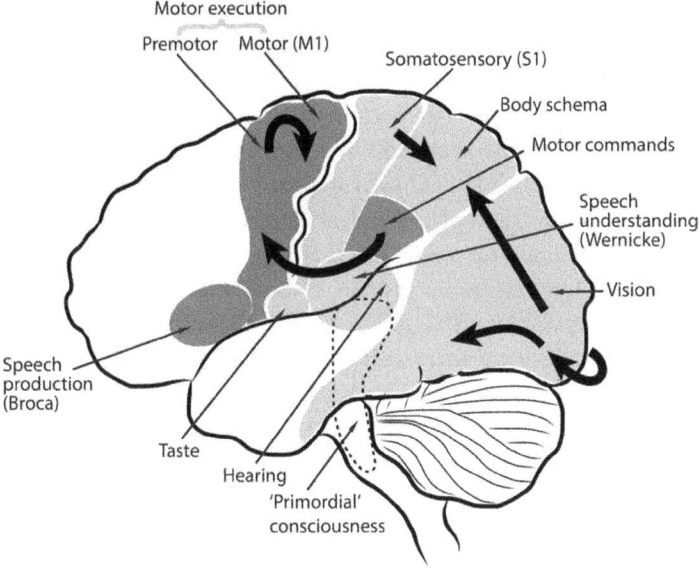

Figure 13.4. The functions mostly associated with different brain areas, including the brain stem. The frontal lobe (including the "prefrontal" region) seems to function as an accessory to the processing and memory areas posteriorly, aiding in judgments and theoretical constructs.

Differences Between Hemispheres

Background

There is one more key aspect of consciousness that needs to be added to those already considered, and this is the differing abilities of the two cerebral hemispheres. That such differences might exist was apparent from the classical neurology and neuropathology of the late 19th century, following Paul Broca's identification of a speech area in the left frontal lobe (in right-handed persons).[29] The observations that really opened up the field, however, came from a very unusual source—patients with "split" brains—and the two neuroscientists who made the greatest contributions were Roger Sperry[30] and Michael Gazzaniga.

At the time that Sperry began his studies on the split brain in the early 1960s, he had already achieved scientific prominence for his pioneering studies on the guidance of growing nerve fibers in the developing nervous system.[31] His chemoaffinity hypothesis postulated that the direction taken by the growing nerve fibers was determined by their ability to respond to chemical signals present in the pathway. The hypothesis was verified by the subsequent discovery of nerve growth factors and led to Nobel Prizes for Rita Levi-Montalcini and Stanley Cohen. Sperry would himself win a Nobel Prize in 1981.

The split-brain work was an entirely different topic from nerve growth, and it originated in an unusual neurosurgical operation, one that involved cutting the corpus callosum, the large bridge of nerve fibers connecting the two cerebral hemispheres. Although there are occasional patients in whom the corpus callosum fails to develop, the likelihood of other congenital neural abnormalities being present or of some functional compensation having occurred through neuroplasticity is high, and neurosurgical patients therefore presented a more satisfactory opportunity for study. The first report of dividing the corpus callosum (callosalectomy) appeared in 1936. Walter Dandy, a pioneering neurosurgeon at Johns Hopkins Hospital, performed the operation to gain better access to the third ventricle. Soon afterwards the operation was used to treat a patient with severe epilepsy, and this last condition was also the reason for a series of callosalectomies undertaken in Los Angeles in the early 1960s. It was these last patients who came to be studied by Roger Sperry and his graduate students.

Results of Sectioning the Corpus Callosum

Although aside from their epileptic symptoms, the treated patients showed little or no effect of the surgery on their daily activities, Sperry demonstrated that appropriate testing could reveal some quite remarkable abnormalities as a result of differences in the functions of the two hemispheres. Of Sperry's graduate students, it was Michael Gazzaniga (Box 13.3) who was mostly responsible for

Box 13.3 Michael Gazzinaga (1939–)

Widely regarded as the greatest living authority on the comparative functions of the two cerebral hemispheres, Michael Gazzinaga was born in Los Angeles in 1939; his father was a surgeon. After high school, Gazzinaga attended Dartmouth College in New Hampshire, gaining a Bachelor in Arts degree in 1961. During his time at Dartmouth he spent a summer with Roger Sperry at the California Institute of Technology in Los Angeles, having been stimulated by Sperry's article in *Scientific American* on "split-brains." Pursuing his interest in this subject, Gazzaniga became one of Sperry's graduate students and was able to examine the first Californian patients to have their corpus callosums divided for intractable epilepsy. Sperry's group had already gathered considerable evidence from animal experiments that the two hemispheres differed in their functional capacities, but the prevailing opinion was that the human brain could be split without interfering with cognition. Gazzaniga, through a variety of tests, showed that this was not the case, since the left hemisphere was far more skilled than the right in interpreting situations and making inferences. After gaining his doctorate, Gazzinaga accepted a teaching position in psychology at the University of California, Santa Barbara, before moving to the New York area where he eventually became director of the Division of Cognitive Neuroscience at Cornell University. While at Cornell, Gazzinaga was able to extend his cognitive studies by examining neurology patients with a variety of mental

Figure B13.3. Michael Gazzinaga, the foremost authority on hemispheric specialization and the founding editor of *The Cognitive Neurosciences*. (Courtesy of Dr. Gazzinaga)

disorders. In 1988 he left New York for Dartmouth College where, apart from four years at the University of California, Davis, he was to remain for the next 18 years. Since 2006 Michael Gazzinaga has been director of the SAGE Center for the Study of Mind at the University of California, Santa Barbara.

In addition to pursuing his own research, Gazzinaga has done much else to develop the area of cognitive neuroscience (including the coining of that term). His activities have included the running of an annual summer school, the founding of a journal, and the editorship of five (massive) volumes of *The Cognitive Neurosciences*. Other publications include a number of books intended for a lay audience, and, outside the laboratory, he has contributed to social, political, and ethical issues of the day.

There is a final remark to be made. The entire body of split brain research could not have been made were it not for the fortuitous conjunction of a novel surgical treatment for epilepsy and the presence of two investigators, Roger Sperry and Michael Gazzinaga, who were available and interested in the possible cognitive consequences of the operation. It is doubtful if more than 50 or so patients were ever treated by callosumectomy and today the operation is hardly, if ever, performed.

The split brain research, with its enormously valuable insights into the comparative functioning of the two hemispheres, is thus another example of chance as a factor in neuroscientific discovery, albeit at a different level to the chance that can occur in the course of an experiment (examples of the latter include Edgar Adrian's detection of the impulse frequency code in sensory neurons and the first recording of miniature end-plate potentials by Paul Fatt and Bernard Katz).

pursuing and extending this line of work. A striking example of the hemispheric difference came from the study of a split-brain patient by Gazzaniga and LeDoux. The experiment involved inserting a screen between the eyes and presenting different pictures to the two eyes; because of the positioning of the pictures the corresponding visual images were referred exclusively to the opposite hemisphere (Figure 13.5). Thus the right hemisphere processed what had been seen by the left eye, and the left hemisphere only dealt with information from the right eye. The subject had a snow scene shown to the left eye and a chicken claw to the right eye. On then facing a set of eight assorted pictures and being asked to choose one related to what had been seen, the subject chose a shovel with the left hand (and right hemisphere) and a chicken with the right hand (and left hemisphere). After

CONTINUING THE SYNTHESIS 249

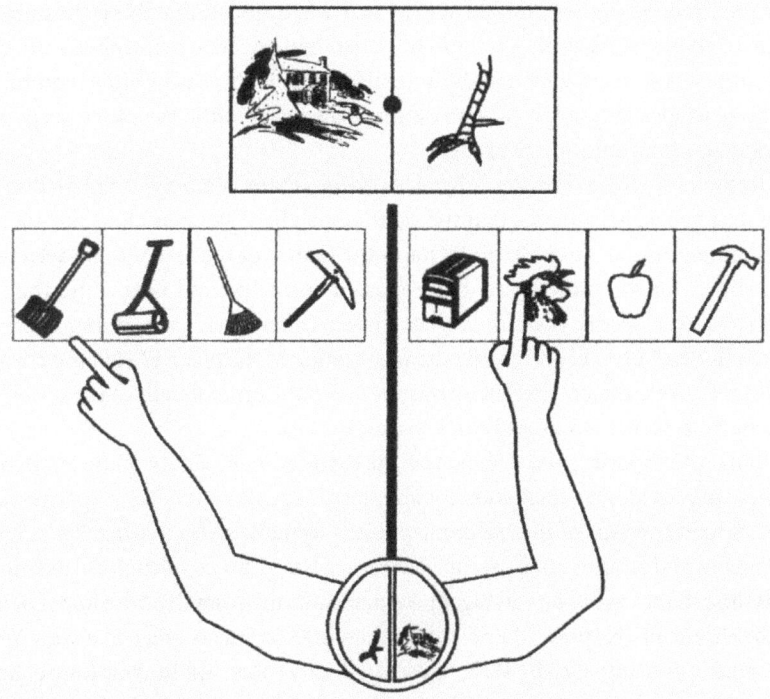

Figure 13.5. Gazzinaga and Le Doux's experiment on a subject with a divided corpus callosum. The left hemisphere "sees" the chicken claw and the right hemisphere sees the snow scene at the top of the figure, but only the left hemisphere can explain subsequent choices made by each hemisphere. (From: Gazzinaga M. Cerebral specialization and interhemispheric communication. Does the corpus callosum enable the human condition? *Brain*. 2000; 123: 1293–1326.)

being asked the reason for this unlikely pairing of objects, the subject replied "Oh, that's simple. The chicken claw goes with the chicken, and you need a shovel to clean out the chicken shed." On further inquiry the subject was adamant in claiming that he had never seen the snow scene.[32]

Interpretation of Results

This dramatic experiment clearly showed that the two hemispheres behaved differently, but it left open the question as to what was happening in the right hemisphere. The simplest answer would be to say that the right hemisphere was unconscious during and after the testing and to extrapolate that this is the normal state of affairs while awake. It could be added that selective hemispheric unconsciousness is seen elsewhere in the animal kingdom—in sea-dwelling mammals, such as whales and dolphins, one hemisphere sleeps while the other keeps the creature swimming and breathing.

In the case of Gazzaniga's subject, however, there is a better explanation of the strange results, and it is that the right hemisphere *was* conscious during the experiment. It did, after all, understand the nature of the test and, by selecting the shovel for the snow scene, had reported correctly. And how could the left hemisphere, responsible for the verbal answer, claim to have seen the snow scene when this had been prevented by the positioning of the picture? Had the corpus callosum been intact, the left hemisphere would undoubtedly have reported seeing both scenes (snow and chicken claw).

What the experiment *did* show was the extraordinary ability of the left hemisphere to provide an explanation for an unusual situation. Indeed, it appears that one of the functions of the left hemisphere is to make sense of whatever is happening in and around one's person. Gazzaniga has given us a delightful example, this time in the case of an intelligent woman suffering from the delusion of being in her home in Freeport, Maine, rather than in Gazzaniga's office in a New York hospital. On being asked: "Well, if you are in Freeport and in your house, how come there are elevators outside the door here?" the grand lady peered at Gazzaniga and replied: "Doctor, do you know how much it cost me to have those put in?" The ingenuity of the brain (presumably the left hemisphere) in creating explanations for improbable actions or beliefs is also seen in hypnosis. On being awoken from a trance, a subject may provide a rational, self-motivated explanation for an irregular physical act that had been suggested by the hypnotist.

The left hemisphere, then, acts as the "interpreter" for the human brain. Various tests in split-brain subjects have shown that, in addition to its language skills, the left hemisphere is better than the right in tackling difficult cognitive tasks. The right hemisphere is far from being a dormant partner, however, since it excels in reading facial expressions and, by using mental imagery, is skilled in solving geometric and other spatial problems. From functional imaging studies, it appears that the right hemisphere also processes music and that its operations may have a strong emotional content. Indeed, Sperry, in his 1981 Nobel Lecture, proposed that, in its rich cognitive and emotional life, the consciousness of the right hemisphere rivaled that of the left.[33] What seems likely, however, is that, depending on the circumstances, one hemisphere may be dominant at any given time. For example, if I am deep in thought and aware of my inner voice as I consider the various issues attached to a difficult problem, perhaps one involving some calculations, it is likely that my left hemisphere is dominating. Nevertheless, if someone unexpectedly taps me on the left shoulder, my right hemisphere is immediately aware of this, showing that, although quiet, it was conscious too. The advantage in having two hemispheres with different specializations is that the brain, by minimizing duplication of function, is able to operate more efficiently.

The Sense of Self

Regarding the sense of self, the testing of split-brain subjects has revealed that, like its partner, the right hemisphere has this quality, being able to recognize pictures of its owner as well as of family and of familiar and historic scenes. Nevertheless, Gazzaniga also suggested that, in its quest to provide explanations for its behaviors, the left hemisphere may make an especially strong contribution to the creation of the sense of "self."

In closing, one cannot do better than to quote Gazzaniga. Though he was speaking of the left hemisphere, the sentiment takes in much of brain function:

> The interpretation of things that we encounter has liberated us from a sense of being determined by our environment; it has created the wonderful sense that our self is in charge of our destiny. All of our everyday success at reasoning through life's data convinces us of this. And because of the interpreter within us, we can drive our automatic brains to greater accomplishment and enjoyment of life.[32]

Indeed!

Notes and References

1. Sherrington CS. *The brain and its mechanism.* Cambridge, England: Cambridge University Press, 1933.
2. Eccles JC. A unitary hypothesis of mind-brain interaction in cerebral cortex. *Proceedings of the Royal Society B.* 1990; 240: 433–451.
3. Libet B. *Mind time: the temporal factor in consciousness.* Cambridge, MA: Harvard University Press, 2004.
4. Huxley TH. On the hypothesis that animals are automata, and its history. *The Fortnightly Review.* 1874; 15 (ns): 555–580. Reprinted in *Method and Results: Essays by Thomas H. Huxley* (New York, NY: Appleton, 1899).
5. Churchland PS. Toward a neurobiology of the mind. In: *The mind-brain continuum: sensory processes.* Llinás R, Churchland PS, eds. Cambridge, MA: MIT Press, 1996:281–303.
6. McGinn C. Can we solve the mind-body problem? *Mind.* 1989; 98: 349–366.
7. Nagel T. What is it like to be a bat? *The Philosophical Review.* 1974; 83(4): 435–450.

8. Spinoza's great work, *Ethics*, which contains the quotation, was published posthumously in Latin in 1677.
9. See Chapter 5 for more information on Sechenov.
10. Watson was especially interested in the mental development of children and made recommendations for their upbringing. Despite his scientific reputation, which led to the presidency of the American Psychological Association in 1916, Watson was obliged to leave his professorial position at Johns Hopkins University because of an affair with a graduate student. His academic career terminated, Watson applied his psychological expertise to the field of advertising and was extremely successful.
11. A good summary of Skinner's deductions concerning human behavior, based on operant conditioning experiments in animals, is given in his books, for example: Skinner BF. *Science and human behavior*. New York, NY: Simon & Schuster, 1953.
12. Walker A. Music and the unconscious. *British Medical Journal.* 1979; 2(6205): 1641–1643.
13. The comment was made in one of Darwin's notebooks and attention has been drawn to it by the late Professor C. U. M. Smith: Smith CU. Darwin's unsolved problem: the place of consciousness in an evolutionary world. *Journal of the History of the Neurosciences.* 2010; 19(2): 105–120.
14. See Chapter 6.
15. The work of Cauller and Kulics, described in Chapter 12.
16. Dade L. How Enigma machines work (enigma.louisedade.co.uk/). For a fuller account, see: Sebag-Montefiore H. *Enigma: the battle for the code*. London, England: Weidenfeld & Nicolson, 2000.
17. Balliet TM. Relation of mind and brain. *The Pennsylvania School Journal.* 1893; 41: 34–35.
18. Bischof M, Bassetti CL. Total dream loss: a distinct neuropsychological dysfunction after bilateral PCA stroke. *Annals of Neurology.* 2004; 56: 583–586.
19. Heron W. The pathology of boredom. *Scientific American.* 1957; 196(1): 52–56.
20. Lilly JC. *The deep self: the tank method of physical isolation*. New York, NY: Simon & Schuster, 1977.
21. Just as in the Charles Bonnet syndrome, in which patients lose their sight, usually because of macular degeneration.
22. McComas A. *The artful chameleon: an exploration of migraine and medicine*. West Flamborough, ON: Alkat Neuroscience, 2006.
23. Though thalamic function will also be disrupted during cortical spike-and-wave activity.
24. Damasio A. *Self comes to mind*. New York, NY: Random House, 2010.
25. An anencephic infant has no, or very little, cerebral cortex, though the brainstem is present. Most infants die within a few days but occasional ones have been kept alive for months and even as long as three years. The study of one such infant is described in Shewmon A. Anencephaly: selected medical aspects. *The Hastings Center Report.* 1988; 18(5): 11–19.

26. Human fetuses also exhibit behavior suggestive of pain (facial grimacing and recoil) if pricked by an amniocentesis needle, as well as of pleasure (thumb sucking), at a time when the cerebral cortex is barely functioning.
27. Head H. *Studies in neurology.* Vols. 1 and 2. London, England: Henry Frowde, 1920.
28. Figure 6.10.
29. See Chapter 6.
30. For more on Sperry, see Box 5.2.
31. See Chapter 5.
32. Gazzinaga MS. Cerebral specialization and interhemispheric communication: Does the corpus callosum enable the human condition? *Brain.* 2000; 123: 1293–1326. (This invited review gives a very readable account of all the split-brain work, including that of the "chicken claw" experiment undertaken with Joseph LeDoux.)
33. Sperry RW. *Some effects of disconnecting the cerebral hemispheres.* Nobel Lecture, December 8, 1981. nobelprize.org.

14
Looking to the Future

Although there are many definitions of consciousness, the one adopted for the purposes of the book is "an organism's awareness of itself." The choice of "organism" is important. On the one hand it leaves space for other animal species, in addition to *H. sapiens*, to have consciousness. In keeping with this possibility, evidence was produced that suggested species as diverse as octopuses and insects had awareness of themselves. On the other hand the definition says nothing about a key issue confronting modern neuroscientists and philosophers—can nonliving systems acquire consciousness? Will robots be able to think for themselves? As others have pointed out, this is a very old question, but the remarkable advances in the capabilities of present-day computers have brought it to the fore.

In considering the question, it is usual to start with Alan Turing (Box 14.1) and his "imitation game," better known as the Turing test.[1] Turing, a self-taught pioneer in machine computing and the person largely responsible for deciphering coded German military messages in World War II, answered the question "Can machines think?" in a practical, if somewhat oblique, way. Turing imagined a situation in which an observer fed questions to a human subject and to a machine, both hidden from the observer. If the (written) answers were such that it was impossible to say whether they had come from the human or the machine, then, logically, was there not an obligation to say that the machine, like the person, had been thinking?

A similar argument grants consciousness not only to modern computers but to any object or system that is able to acquire information about itself and its surroundings and, on the basis of that information, undertake actions judged appropriate by a human observer. In this case, a multitude of devices would qualify—a thermostatically controlled heating system, a driverless car, and an automatic domestic vacuum cleaner would be three obvious examples. Although such a position may sound extreme, there are thoughtful and well-informed scientist-philosophers who are prepared to go very much further, even to the point of considering elementary consciousness to be an intrinsic property of all matter. For example, in a radio discussion on consciousness, David Chalmers, a leading expert on the "hard problem," stated seriously but with some humor:

> Even a photon has some degree of consciousness. The idea is not that photons are intelligent or thinking. It's not that a photon is wracked with angst because

Box 14.1 Alan Turing (1912–1954)

Alan Turing, a pioneer in AI and computer design, was a largely self-taught mathematical genius who played a pivotal role in World War II by deciphering German secret codes. Neither of Turing's parents had backgrounds in science: his father held a senior position in the Indian civil service and his mother enjoyed the privileged life of a young British woman in India. Turing was placed in a private boarding school in England where, after a poor start, he eventually thrived, astonishing his teachers with the originality and scope of his mathematics. Turing continued in mathematics after winning a scholarship to Cambridge where he graduated with distinction and subsequently became a Fellow at Kings College. It was there that he met the leading mathematician John von Neumann, visiting from Princeton, New Jersey, and, probably as a result, came up with the idea of a computing machine. It was a concept that, among many potential applications, was highly relevant to Turing's longstanding interest in the nature of the human mind. As he envisaged it, a machine, operated by algorithms, would be able to perform all the functions that a human "computer" was capable of. Turing used the concept as the basis

Figure B14.1. Alan Turing; a passport photograph taken at 16. (Wikimedia Commons)

> of a proof concerning computable numbers, a very considerable achievement that led to his becoming a Visiting Fellow at Princeton and a PhD.
>
> After World War II broke out Turing took a leadership role in deciphering the German secret messages encrypted by that country's formidable Enigma machine (see Chapter 13). Turing solved the problem by conceiving and helping to design electromechanical machines ("bombes") that operated by following logical rules. This success, a highly guarded secret, was of vital importance, particularly in overcoming the menace of the German U-boats to Allied shipping. After the war, Turing, while working at the National Physical Laboratory, produced the first detailed design of a stored-program computer, and he continued designing computers after moving to the Victoria University of Manchester. While in Manchester he published the influential paper on machine intelligence that included the proposal for an "imitation game" (subsequently known as the Turing test). Turing's broad interest in science was then reflected in a totally new direction—the application of mathematics to explaining the development of patterns and shapes during embryological development.
>
> Sadly, this gifted life came to an end not long after a conviction for indecent behavior (Turing was homosexual); whether the self-inflicted cyanide poisoning was deliberate or accidental remains moot. In later years, Alan Turing's achievements have been recognized in innumerable ways throughout the world. The biography by Andrew Hodges (*Alan Turing: The Enigma of Intelligence* [London, England: Burnett Books, 1983]) is superb. The 2014 film, *The Imitation Game*, with Benedict Cumberbatch as Turing is based on the latter's wartime work at Bletchley Park.

it's thinking, "Aww, I'm always buzzing around near the speed of light. I never get to slow down and smell the roses." No, not like that. But the thought is maybe photons might have some element of raw, subjective feeling, some primitive precursor to consciousness.[2]

The mention of "matter" raises a further point favoring the potential for consciousness in inanimate objects. Thus all forms of life, like everything else in the universe, are composed of chemical elements. Moreover, the cell metabolism that drives all the physiological body systems, including the ion pumps responsible for the membrane potentials of excitable cells, are ultimately physicochemical reactions. Why should it not be possible, using the same elements, to simulate or re-create these reactions and, in so doing, bring about consciousness?

Moving up in complexity, from molecules to cells to creatures, consider also a situation in which nerve fibers are removed, one by one, from a conscious organism and replaced in each case by a length of fine insulated wire. A wire would be able to conduct electric signals as well—indeed better—than a nerve cell. How, then, could consciousness not persist in a nervous system that was now composed largely of inorganic material?

At first glance the wire substitution scenario, which has been raised elsewhere, seems a powerful argument for inanimate consciousness, but closer examination suggests otherwise. Though the conduction of impulses along the axon of a neuron is a relatively simple task, the impulse must then invade the many branches of that axon, exerting excitatory or inhibitory influences at the multitude of synapses with tens or even hundreds of other cells. It would, in effect, be a technical impossibility to replace the fine architecture of the nervous system with electronic components.

Suppose, however, that one were to start from scratch by designing and building an entire artificial nervous system using electronic components. This is something that was done in the 1940s by William Grey Walter (Box 14.2), one of the pioneers of EEG, at a time when there were no transistors or integrated circuits available and only resistors, capacitors, valves, and the like. Grey Walter's creation was a pair of mobile devices that, with their detachable covers in place, resembled tortoises. Each "tortoise" had a rotating photocell that, with appropriate circuitry, directed the robot to the strongest light source in its neighborhood; motion was effected by a motor driving the two rear wheels with a single wheel at the front for steering (Box 14.2). If the tortoise encountered an object in its path, touch sensors would cause it to halt and try an alternative route. Importantly, when its battery power was low the tortoise would, without any outside prompting, make its own way to the charging station in its kennel. So here was a device that, like a present-day driverless car, was able to sense information about itself and its surroundings and act appropriately. Modestly, Grey Walter claimed that what he had created was equivalent to a nervous system with only two cells.

To an extraterrestrial visitor, the activity of the tortoise appears as intelligent behavior little different from that of many "simple" animal species. To some neuroscientist-philosophers the tortoise, by virtue of that seemingly intelligent behavior, possesses a consciousness, as do the many varieties of modern robots, at least some of which are now able to interpret and respond to human facial expressions. However, even if one ignores the objection that behavior deemed to be intelligent does not necessarily denote consciousness, there is still the further issue that all the activities of a robot are dependent on human intervention—either in designing and building electronic circuits, as in the case of Grey Walter's tortoises, or in writing computer programs for the robots of the present time. Leaving aside the counter argument, that a manmade circuit or program is conceptually equivalent to a system of neural pathways constructed from a genetic

Box 14.2 William Grey Walter (1910–1977)

Grey Walter was born in Kansas City, Missouri, but was educated in England, ultimately attending King's College, Cambridge. Attracted to neurophysiology, he would, with his intelligence and practical gifts, have seemed an automatic choice for a research fellowship under Adrian in the Cambridge Physiological Laboratory. Instead he was obliged to take positions in London hospitals before being offered a more prestigious one at the Burden Neurological Institute in Bristol. By this time his interest was mainly in the EEG, and he had not only visited Hans Berger, the German discoverer of the EEG, but had, like Dawson, built his own versions of EEG machines. During World War II Walter, as a scientist, was active in radar and guided missile research. Returning after the war to the Burden Neurological Institute, Walter

Figure B14.2. William Grey Walter operating on the rear of Elsie, one of his light-seeking tortoises, in 1950. The rotating photocell is sticking up at the front, above the front wheel used for steering; a "shell" (in the background) normally covered the motor and circuitry. (Reproduced from Moravec H. Robot. *Mere machine to transcendent mind.* New York, NY: Oxford University Press, 1998.)

was able to make recordings with intracerebral wire electrodes in patients being considered for neurosurgery.

An inventor, Walter devised a "toposcope"—an array of small cathode ray tubes, each of which registered the amount of EEG activity in a given region of cerebral cortex. Perhaps more important, and certainly more fascinating, was his earlier creation of two robots in 1948–1949; these three-wheeled machines came to be known as "tortoises," partly due to their slow movements. Even though the electric circuitry involved was relatively simple, the tortoises were able to make their way, when necessary, to a battery recharging station. One of the tortoises was subsequently modified to be able to exhibit what, in a live animal, would be regarded as conditioned reflex behavior. In these early endeavors at AI Walter emphasized the importance of using analogue rather than digital circuitry. As far as the human brain was concerned, however, Walter's most important discovery was probably the detection of the "contingent negative variation" (see main text).

An important and influential scientist, Walter was also known for his left-wing views. His research career was compromised by a brain injury, incurred in a motor scooter accident when he was 60, seven years before his death.

blueprint, is it possible to create an artificial nervous system that resembles a naturally occurring one in its ability to learn?

For those in the field of artificial intelligence (AI), the answer to this question is an emphatic "yes."[3] Indeed, it is this extra operation of the machine, its "learning" ability, that distinguishes a search engine such as Google from a captive information device like IBM's "Watson." While the latter could respond to a human voice and win at Jeopardy, just as IBM's "Deep Blue" could defeat the world chess champion, it was only because of the facts and rules made available to the respective computers by their human programmers.

It was therefore a major advance when, using a learning approach, Google's AlphaGo was able to defeat the world's greatest GO experts, and do so convincingly. As complex as chess is, GO is more so—the number of possible moves greatly exceeds the number of atoms in the universe! However, to more closely resemble the way a brain works, a machine would have to learn from the input data by itself, without any human instructions. One of the pioneers of this "neural net" approach is George Hinton (Figure 14.1) now a driving force with Google Brain, who pursued his ideas during doctoral studies at Edinburgh University in the 1970s.[4] By building multiple layers of artificial neuron nets, a machine can learn to recognize patterns in the information available and then discern

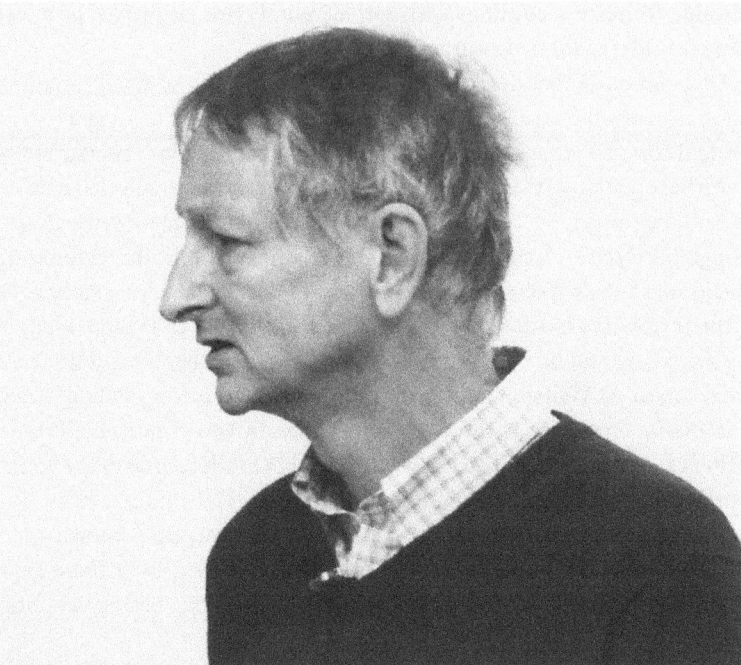

Figure 14.1. Geoffrey Hinton (1947–). A professor at the University of Toronto, Hinton is a leader in AI and a great-great grandson of the mathematician and logician George Boole. (Photo by Eviatar Bach, Wikimedia Commons CC BY-SA 3.0)

patterns of patterns—thereby operating much as Konorski had conceived his sensory analyzers to work in living brains, with their creation of gnostic units. Further, Hinton's use of back-projections also mirrors the situations found for vision, hearing, and touch described earlier.

The neural net approach has now succeeded, both for vision and for language translation. In the case of the former, a learning process entailed the machine[5] being presented with 10 million frames from unselected videos posted on the Internet's YouTube. The machine was then able to construct the image of a cat's face and to recognize the faces of cats in assorted pictures; similar success was achieved with images of human faces and bodies.[6] However, although the neural net approach is clearly the way for AI to develop, it still suffers from inefficiency and haphazard outcomes, so different from the purposeful learning of a living creature. What is needed, surely, is some kind of potential reward for the machine. For Grey Walter's tortoise, its ability to recharge its own battery could be considered a reward for having conducted explorations of its surroundings, and perhaps the delivery of an electrical charge to the operating system might similarly assist computer-based learning. And, again, if consciousness is considered not only as an ability to learn but also a self-awareness on the part of the operating system, might the latter not be facilitated by the possession of an external form, as in the artificial shells of Grey Walter's tortoises?

At this point it is instructive to examine an extension to a very challenging line of thought in the field of nonhuman intelligence and consciousness—the "Chinese Room" argument[7] devised by the contemporary philosopher John Searle (1932–). Searle imagined himself locked in a room and being given written messages. Having been provided with a book of instructions, in English, he is able to write a response in a way prescribed for each new message. To observers outside the room, unaware of the book of instructions, the responses sent out are evidence of Searle's intelligence. The catch in the story is that he, Searle, understands nothing of the information being exchanged—it is all written in Chinese characters! In this imagined situation Searle is acting just as a computer program would, and, since he is unaware of the content of the messaging, so would a computer be. The modification to the story is that the argument would still stand if Searle became sufficiently familiar with the Chinese characters that he could dispense with the instruction book altogether. However, though learning had occurred, as in an AI machine, it was "syntactical" (concerned with procedure) rather than "semantic" (concerned with meaning). Thus the behavior, which had previously seemed intelligent, was nothing of the kind. Because of the power of Searle's line of thought, it is not surprising that his argument has been discussed repeatedly by philosophers ever since; Searle himself has subsequently responded to some of the criticisms, though without taking up the issue of machine learning.

Imagine now a still more demanding type of scenario, one that makes it even more unlikely that machine will ever acquire the consciousness necessary for experiencing a "qualia"—that is, a particular color, sound, touch, or smell (the "hard problem"). One of the first examples is that provided by the English philosopher C. D. Broad almost a century ago. Using the sense of smell for his argument, Broad posited the existence of a scientifically gifted archangel:

> He (the archangel) would know exactly what the microscopic structure of ammonia must be; but he would be totally unable to predict that a substance with this structure must smell as ammonia does when it gets into the human nose. The utmost he could predict on this subject would be that certain changes would take place in the mucous membrane, the olfactory nerves and so on. But he could not possibly know that these changes would be accompanied by the appearance of a smell in general or of the peculiar smell of ammonia in particular, unless someone told him so or he had smelled it for himself.[8]

A more modern version of this paradigm, and one especially suited to neurophysiologists since it concerns one of their number, is that devised by Frank Jackson (1982):

> Mary is a brilliant scientist who is, for whatever reason, forced to investigate the world from a black and white room via a black and white television monitor. She specializes in the neurophysiology of vision and acquires, let us suppose, all the physical information there is to obtain about what goes on when we see ripe tomatoes, or the sky, and use terms like "red," "blue," and so on. She discovers, for example, just which wavelength combinations from the sky stimulate the retina, and exactly how this produces via the central nervous system the contraction of the vocal cords and expulsion of air from the lungs that results in the uttering of the sentence "The sky is blue . . ." What will happen when Mary is released from her black and white room or is given a color television monitor? Will she learn anything or not? It seems just obvious that she will learn something about the world and our visual experience of it. But then is it inescapable that her previous knowledge was incomplete. But she had all the physical information. Ergo there is more to have than that, and Physicalism is false.[9]

Though commonly used, this argument seems unconvincing. Despite her studies, Mary evidently did *not* know everything about vision; otherwise she would have been aware that there were some individuals who were born with total color-blindness. Indeed, in her reading and conversations, Mary must have been puzzled by mention of the beauty of blue skies and red sunsets and may well have wondered if she was one of those born without color vision. (A possible objection to the original conundrum and to its addendum is that two entirely different types of information, sensory and semantic, are being mixed together.)

While an argument such as Broad's make it doubtful whether machines will ever develop consciousness, there are those, such as the noted AI expert Ray Kurzweil, who are prepared to make "a leap of faith." Extending the "imitation" argument of Alan Turing, Kurzweil has written:

> My objective prediction is that machines in the future will appear to be conscious and that they will be convincing to biological people when they speak of their qualia. They will exhibit the full range of subtle, familiar emotional cues; they will make us laugh and cry; and they will get mad at us if we say that we don't believe that they are conscious. (They will be very smart, so we won't want that to happen.) We will come to accept that they are conscious persons. My own leap of faith is this: Once machines do succeed in being convincing when they speak of heir qualia and conscious experiences, they will indeed constitute conscious persons.[10]

Perhaps—but, like our inability to "know" what it is like to be a bat, we humans will never know what it is like to be one of Kurzweil's machines. Whether it is truly conscious is likely to remain unanswerable.

Notes and References

1. Turing AM. I. Computing machinery and intelligence. *Mind.* 1950; 59: 433–460. This is a very interesting paper and is written in a way that makes it easy to understand. It provides an introduction to digital computing and continually compares the operations and abilities of a computer to those of a human mind. Turing regarded the question "Can machines think?" as meaningless, proposing instead that machines should be judged by what they could do. This, in essence, was the basis of the Turing test, a type of imitation game. Turing foresaw that future machines would be able to learn.
2. Chalmers D. How can we explain the mystery of consciousness? TED Radio Hour Discussion, February 20, 2015. npr.org.
3. Lewis-Kraus G. The great A.I. awakening. *New York Times Magazine*, December 14, 2016. This lively article is not only informative about recent advances in computer learning but captures the personalities and working ambience of some of those engaged in the field.
4. Geoffrey Hinton (1947–) is a great-great-grandson of the mathematician and logician George Boole. Hinton was born in Wimbledon, United Kingdom, and gained a bachelor of arts in experimental psychology at the University of Cambridge. Next came doctoral studies in AI at the University of Edinburgh, this at a time when the topic was still in its infancy. As a result of his pioneering work, Hinton has received numerous honors. He is presently an emeritus distinguished professor at the University of Toronto and also directs the Toronto branch of Google Brain.
5. Actually 1,000 computers operating as a single unit.
6. Le QV, Ranzato M, Monga R, et al. Building high-level features using large scale unsupervised learning. *Proceedings of the 29th International Conference on Machine Learning*. Edinburgh, England, 2012.
7. Searle J. Minds, brains and programs. *Behavioral and Brain Sciences.* 1980; 3: 417–424.
8. Broad CD. *The mind and its place in nature*. New York, NY: Humanities Press, 1925.
9. Jackson F. Epiphenomenal qualia. *Philosophical Quarterly.* 1982; 32: 127–136.
10. Kurzweil R. *How to create a mind: the secret of human thought revealed*. New York, NY: Penguin, 2012. Kurzweil, born in New York in 1948, is presently Google's director of engineering. A brilliant inventor (e.g., flatbed scanner, text-to-speech synthesizer, music synthesizer), Kurzweil is known for the success of his predictions regarding the place of science in the world.

15
Summary

Having assimilated the key observations to do with consciousness (and those responsible for making them), it is time to bring them together in a summary—a summary that includes much of the speculative material in the last two chapters. Here it is then:

(i) In all species that possess it, consciousness is entirely the product of electrical events in nerve cells.

(ii) It follows that "free will" in humans is an illusion, the illusion made possible by the enormous numbers, rich variety, and interactions of "gnostic units" (neural assemblies coding for people, places, objects, actions, etc.). Adding to the illusion of free will is the ability of the left hemisphere to rationalize actions undertaken by an individual.

(iii) In humans, different regions of cerebral cortex, working in conjunction with the thalamus, are responsible for different types of consciousness. The most important of these regions is at the back of the brain, for it is vision, with its huge cortical representation, that drives us. It is what we see at the moment, or what we can imagine taking place, that usually determines what we will do next.

(iv) The visual system does more, however—together with information from receptors on and below the body surface, it creates an unconscious "schema" of the whole person, a neural representation that is being continually updated.

(v) All the while, accompanying our actions, is an inner voice. The voice is the product of gnostic unit activity in the speech area; it is a voice that provides a commentary throughout the waking day. And further enriching our conscious experience are other cortical sensory areas, such as those for touch, smell, and taste.

(vi) At the very front of the brain are the neurons that have been requisitioned by the more posterior areas to assist and further their activities. It is in this prefrontal cortex that plans are made and the rich variety of abstract thinking takes place.

(vii) In all the cortical transactions, motor and sensory, the fundamental executive units are columns of neurons that, together with their axons and dendrites, are aligned perpendicular to the surface of the brain. There is a plethora of these columns, as many as 20 million in the human brain, and only a fraction of them are used in the lifetime of the individual.

(viii) In the awake state the columns fluctuate in their excitability and responsiveness, doing so independently of neighboring columns; the moment-to-moment excitability is governed by one or more pyramidal cells deep in the column, through inhibitory interneurons.
(ix) The columns form the successive hierarchies in a sensory system, culminating in "gnostic units" specific for persons and other visual objects, as well as for sounds, tastes, smells, and touches. The gnostic units involving the body enable the organism to create an unconscious representation ("body schema") of itself, while the convergence of different types of gnostic units creates "concept" ("grandmother") units and memories of people, objects, and events. The recall of a memory, the ability to imagine and to dream, all depend on back-projections from the concept cells to the cortical sensory areas.
(x) A different type of gnostic unit—the kinesthetic unit—is most prevalent in the parietal lobe. These units provide the earliest preparation for movement, a preparation that is continued by more anteriorly situated neurons around the primary motor area.
(xi) In the sensory areas and probably in other areas of cortex too, the activity in the cortical columns is "read out" every 30 to 100 ms since, rather than running continuously, the cortex operates in small "chunks" of time. Within these chunks, the high impulse firing frequencies of the pyramidal cells in the columns result in conscious perceptions and thoughts.
(xii) In a sensory system the back-projections to a primary cortical receiving area take place within a time-chunk, with the back-projections summing with the synaptic activity still present in the apical dendrites of the pyramidal cells. It is likely that the back-projections provide the detail in the sensory experience, the nature of the experience having been determined by gnostic units higher in the sensory pathway.
(xiii) If, one by one, the cortical areas and their columns fall silent, there will still be a primordial consciousness, a vague awareness tempered by pleasure and pain. It is a consciousness that is present in a newborn infant, at a time when behavior is entirely driven by reflexes. That there should be this basic awareness is because of the activities of the limbic system and brain stem.
(xiv) The future development of consciousness in a nonorganic system can neither be discounted nor proven to have taken place.

Evidence

The strength of the evidence for each of the preceding statements varies. On one hand, some propositions are based merely on the supposition that it "would make sense" if an anatomical or physiological feature functioned in a certain way (such as sensory processing by back-projections). On the other hand, the proposition for "time-chunking" seems irrefutable in view of the abundant evidence from masking experiments. Further, given the existence of time-chunking, then not only is the case for "binding" of sensory features by common rhythmical activity untenable but one can look for neurophysiological activity that would fit in with time-chunking, such as the burst discharges of the layer V pyramidal cells (and, at the same time, consider alternative functions for the rhythmical activity). Table 15.1 lists some of the main propositions of the book and indicates the source and weight of evidence for each one.

Table 15.1

Strength of Evidence	Proposition	Evidence
*****	Back-projection to sensory cortex	Anatomy, latencies, functional magnetic resonance imaging
*	Significance of back-projections	Simplest explanation of vision, touch. etc.
*****	Time-chunks	Masking experiments
*	"Burst" pyramidal cells as consciousness generators	Fits in with time-chunking
*****	"Gnostic units"	Recordings from "face," "object" and concept cells
*****	Categories of gnostic unit	Neurological patients, introspection
***	Cortical columns as gnostic units	Necessity; sensory features similar in cells within column, structure of column
****	Intracolumnar dynamics	Neuroanatomy and cell recordings
*	Autonomy of cortical columns	Noncorrelated spontaneous firing.
*****	Surplus of columns	Minimal effects of prefrontal lesions; savant achievements; nonresponsive spontaneously active cells
*	Cortical and thalamocortical rhythms as neural "noise" suppressors	Long durations of inhibitory postsynaptic potentials i following subthreshold excitatory postsynaptic potentials; brain observations (Llinás)

* = weak; ***** = very strong.

16
Postscript

Charles Sherrington's masterful imagery of the brain began the book. He likened the brain's activities to "little points of light," of which "some stationary flashed rhythmically" while others streamed "in serial trains at various speeds" to junctions where they diverged. While this served as a good analogy for the impulses in the nerve fibers, something more was required for the transition of the sleeping brain into wakefulness—hence the allusion to "an enchanted loom where millions of flashing shuttles weave a dissolving pattern, always a meaningful pattern though never an abiding one."

In the many decades that have passed since the publication of Sherrington's Gifford Lectures, we have come to know much more about the workings of the brain and the underpinnings of consciousness. We can now, for example, replace the image of the loom with columns of cortical neurons as they become, either transiently or permanently, Konorski's gnostic units. Doubtless more detail will emerge concerning the brain's other activities but, as the book has attempted to show, a plausible account of the background to consciousness can now be given from the key observations already made.

The choice of Sherrington's words reflects admiration for his mastery of philosophy and poetry. As a neurophysiologist, even as a physiologist, Sherrington was rated by his contemporaries as the greatest of his time. Looking back now, however, one might put Edgar Adrian ahead of Sherrington, at least in terms of pioneering discoveries; both of them were to share the Nobel Prize in Physiology or Medicine in 1932. And then there was John Eccles, who emerged from Sherrington's department in Oxford to set up the dominant neurophysiological laboratory in the newly created Australian National University. No part of the nervous system was beyond the reach of Eccles' microelectrodes and his rigorous analysis of the recordings; little wonder that so many established neurophysiologists came to spend time with him in Canberra.

But there was one man who, by describing the structure of the nervous system in such exquisite detail and correctly suggesting some of its functions, may have been the greatest neuroscientist of all—Ramon y Cajal. Then again, if, as in chess competition, one were to award a prize for brilliance, how could one decide between the claims of Hermann Helmholtz, Julius Bernstein, Andrew Huxley, Roderick Mackinnon, and many others?

It is a sad thing that some of these names and many others in this book will be unfamiliar to new generations of neuroscientists. But such is the way of the world, for as a wise man observed six centuries ago:

> Where are they now, Doctor This and Professor That, whom you used to hear so much about when they were alive, and at the height of their reputation? They have handed over their chairs to other men, who probably never waste a thought on them; while they lived, they counted for something, now they are never mentioned.[1]

To which a poet, emulating Sherrington, might add:

> And yet 'tis right that
> Neural men[2] still left
> With furrowed brow and fingers deft
> Might momentary pause
> And once again admire
> The heights to which their
> Late lived masters
> Did earnestly aspire.

Notes and References

1. St. Thomas à Kempis (ca. 1418). *The Imitation of Christ*. Translated from the Latin by Ronald Knox and Michael Oakley. London, England: Burns & Oates, 1959.
2. And "neural women" too!

Acknowledgments

There are two people and two institutions to whom and to which I owe an enormous debt.

But before I get to them there are a number of people who deserve a heartfelt thank-you. In no particular order these include Whitney Rauenhorst (University of Chicago Press), Philip Skroska (Becker Medical Library, Washington University), Jack Eckert (Francis A. Countway Library of Medicine, Harvard University), Bill Harris (Department of Physiology, Development and Neuroscience, Cambridge University), Jack McMahan (Stanford University), Sandra Witelson (McMaster University), the late David Hubel (Harvard University), Michael Gazzinaga (University of California, Santa Barbara), Igor Timofeev (Laval University), Horace Barlow (Cambridge University), Rodrigo Quiroga (University of Leicester), and Lincoln Stoller—all of whom allowed or helped me to obtain material for reproduction. It was a joy to receive their positive replies. I also gained useful information and suggestions from the late Douglas Stuart (University of Arizona) and from unknown reviewers of the manuscript.

And then there were and are my friends—Lud Prevec, George Sweeney, Karl Freeman (now deceased), Erwin Montgomery—all of whom encouraged me during our varied social interactions. Jean Delbeke (Brussels) not only encouraged me but provided valuable criticism of the manuscript. Going back further, Adrian Upton has been a colleague at McMaster University for almost 50 years, and Roberto Sica (University of Buenos Aires) is not only a very close friend but was always the "go-to" person when good photos were needed for a paper. I grieve over Roberto's illness.

It would be a terrible omission not to thank Oxford University Press for taking on this book, and I am especially indebted to Joan Bossert and Philip Velinov for their editorial skills and many helpful suggestions. Nearer home, Aurelia Shaw proved invaluable in the correction of proofs.

Now for the institutions—the Wellcome Foundation and Wikimedia Commons. Without the availability of their copyright-free images this book would have been a rather sorry affair—plenty of words but short on illustrations to spice up the text. The Wellcome Foundation was familiar to me during the writing of *Galvani's Spark*, but Wikimedia Commons was a fresh discovery and a very important one at that. I cannot adequately thank those who created it and have supplied material used in this book, and on behalf of authors like myself,

I wish the Commons continued success. In sharp contrast to that pleasant situation, it is shameful that authors now have to pay substantial fees for reproducing material from journals and books, and all credit to the few publishers who are exceptions!

I must also gratefully acknowledge two highly important sources of information. One is Google—no surprise there!—and the other is the ongoing series of illuminating life essays written for *The History of Neuroscience in Autobiography*, edited by Larry Squire (Elsevier). The essays make absorbing reading and the best create vivid mental pictures of the neuroscientists, their achievements, and their circumstances.

Lastly, but most important of all, my thanks go to the two Maries. One of them, Marie Levesque, came to the rescue when fresh illustrations were needed and exhibited not only a great talent but also extraordinary forbearance when changes were constantly being made. And always so pleasant, too!

The other Marie is my wife, Marie Ambruz, to whom the book is dedicated. It was this Marie who rescued me from a bleak existence and has given me six of the happiest years of my life—as well as a very comfortable study in which to write. Thank you, my love.

Index

Note: Tables, figures, and boxes are indicated by *t*, *f*, and *b* following the page number

For the benefit of digital users, indexed terms that span two pages (e.g., 52–53) may, on occasion, appear on only one of those pages.

Adrian, Edgar
   biography, 129*b*
   demonstration of nerve impulses, 60–61
   photograph, 106*f*
   plasticity and mapping of the brain, 94
   on wakefulness of cerebral cortex, 115
   work on electricity and the brain, 128–31
Albe-Fessard, Denise, intracellular recordings, 134–35
"Alex," African Grey parrot, 38–39
Amassian, Vahé
   biography, 180*b*
   photograph, 180*f*
   "time chunk" experiments, 179–81
Amedi, Amir, research on plasticity, 97–98
Amenhotep, physician of ancient Egypt, 12
American Sign Language, chimpanzee study of, 41–42
Andersen, Per, thalamic nuclei, 134
Animal Minds (Griffin), 38
animals, consciousness in
   availability of neurons for, 44–47, 45*f*, 46*t*
   Cambridge Declaration on Consciousness, 47
   Charles Darwin and *The Descent of Man,* 32–36
   crows, 39, 40*f*
   dogs and cats, 40
   dolphins, 42–43
   historical attitudes toward, 32, 47n1
   insects, 43*f*, 43–44
   neuroanatomy and neurophysiology, 45–47, 46*t*
   nonhuman primates, 41–42
   octopi, 40–41
   parrots, 38–39
   and problem-solving in animals, 37–38
   T. H. Huxley on, 36
Animal Thinking (Griffin), 38

artificial intelligence
   neural net approach to building, 259–61
artificial intelligence, learning ability of, 257–61
artificial nervous systems, design and building of, 257–59
assassin bugs, complex behavior of, 43*f*, 43
astrocytes, role among glial cells, 59
auditory system
   acoustic analyzers, 201–3
   and recovery from deafness, 178
Awakenings (Sacks), 122n27
"Ayumu," chimpanzee at Kyoto University, 42

Bach y Rita, Paul
   research on plasticity, 97–98
back-projections
   advantages of, 240
   in dreams and imagination, 240–41
   and function of gnostic units, 183
   and sensory experience, 237
backward masking, 179–81, 182*f*
Balliet, T. M., back-projections in dreams and imagination, 240–41
Bartholow, Roberts
   motor cortex, mapping of, 76
Bartley, Howard
   visual cortex, mapping of, 79–82
bees, problem-solving among, 38, 43
behavior, reflexive nature of, 229–36
behaviorism, school of psychology, 232
Bereitschaftspotential (readiness potential), 189–92, 191*f*
   commentary on, 195–98
   cortical processing time, 196
   free will experiment, 195*f*, 196–97, 228–29
   intended movement, ability to cancel, 197–98
   simplicity of tasks studied, 198
   supplementary motor area (SMA), 197
   visual processing time, 197

Berger, Hans
    biography, 127*b*
    and electroencephalogram (EEG), 126–28
    equipment used by, 126
    portrait, 127*f*
Berger rhythm, 94
Bernstein, Julius
    demonstration of nerve impulses,
        60–61, 62*f*
    photograph, 61*f*
Beth Abraham Hospital, New York
    and "encephalitis lethargica," 122n27
biography, of Charles Scott
    Sherrington, xiv–xv
birds, consciousness and cognitive abilities
    in, 38–39, 40*f*
Bishop, George
    investigation of slow waves as synaptic
        potentials, 131
    life and research, 131, 139n9
    portrait, 132*f*
    visual cortex, mapping of, 79–82
bison, 19th-century slaughter of, 47n1
Blakemore, Colin
    biography, 85*b*
    fetal cortex, development of, 94–97
    on neuroplasticity of fetal cortex, 236–37
    photograph, 95*f*
blindness, studies of recovery from, 177–78
Bliss, Timothy
    creation of memory, 92
body schema
    clinical distortions of, 88
    loss of, 243–44
    and somatosensory cortex, 87
Bonnet, Charles, syndrome named for,
    224–25n23
brain
    consciousness and patterns of activity, 138
    and debate on site of consciousness,
        14–21, 29–30
    differences between hemispheres, 246–51
    discovery of electric impulses in, 12–14
    early writings on, 12–14
    lateral surface of left cerebral
        hemisphere, 15*f*
    neurotransmitters in, 115–18
    regions of, 245*f*
brain, and electricity
    and electroencephalogram (EEG),
        demonstration of, 131

    and electroencephalogram (EEG),
        development of, 126–28
    and electroencephalogram (EEG),
        recordings of, 130*f*
    *A History of the Electrical Activity of the
        Brain,* 139n4
    impulses *vs.* slow waves, 123–26
    interactions between thalamus and
        cortex, 136–38
    intracellular recordings from
        motoneurons, 131–33, 133*f*, 139n12
    readiness potential
        (Bereitschaftspotential), 189–92, 195–98
    signal averaging, development of, 186–89
    single cell studies in cortex, 134–36
    slow waves as synaptic potentials, 131–34
    work of Edgar Adrian at Cambridge
        laboratories, 128–31
brain, electricity and
    glass microelectrodes, use of, 141
brain, mapping of
    cerebral cortex, role in consciousness, 69
    consciousness, relationship of motor and
        sensory areas to, 90–91
    cortical localization, 70
    external features of the brain, 69
    hearing, taste, and smell, location of, 90
    Human Brain Connectome Project, 94
    long-term memory, 92–93
    motor cortex, 71–77
    plasticity and, 94–98, 97*f*
    short-term memory, 91–92
    somatosensory cortex, 82–90
    visual cortex, 78–82
brain, neuroanatomy and
        neurophysiology of, 49
    glial cells, 59–60
    glial cells, average number of, 66–67
    nerve impulses, 60–63
    neurons, 49–59, 52*f*, 53*f*
    neurons, average number of, 66–67
    perspectives on research, 67
    synapse, excitation and inhibition at
        the, 63–66
brain, synthesis of activity, 226
    aims and limitations of, 227–28
    back-projections, advantages of, 240
    back-projections, and sensory
        experience, 237
    back-projections, in dreams and
        imagination, 240–41

body schema, loss of, 243–44
brain regions, 245f
brain regions and contributions to
   consciousness, 242–45
and concept of free will, 228–29
corpus callosum, results of sectioning,
   246–51, 249f
creativity, unexpected nature of, 235–36
differences between hemispheres, 246–51
dualism and monism, 226–27
Enigma Machine, analogy of, 237–40
and evolution of sensory
   consciousness, 241–42
pain, ability to experience, 244–45
pleasure, ability to experience, 244–45
reflexive nature of human
   behavior, 229–36
sense of self, 251
sensory deprivation, effects of, 243
sensory receptors, and specifying cortical
   modality, 236–37
brainstem
   and debate on site of
      consciousness, 14–15, 15f, 29–30
   function of, 103–5
Brazier, Mary, *A History of the Electrical
   Activity of the Brain*, 139n4
Bremer, Frédéric
   biography, 104b
   life and research, 103–5, 119–20
   reticular activating system, 115
Broad, C. D., nonhuman consciousness, 261
Broca, Pierre Paul
   cortical localization, 70
   portrait, 70f
Brock, Lawrence, slow waves as synaptic
   potentials, 132–33
Brodmann, Korbinian
   activity within cortical columns, 210
   areas of cerebral cortex, 80f
   portrait, 79f
   visual cortex, mapping of, 78–79
Burns, Delisle, cortical structure, 212–13
Buser, Pierre, intracellular recordings, 134–35

Cajal, Santiago Ramón y
   anatomical studies of retina, 161–62, 162f
   contributions of, 267
   life and research of, 49–53, 67n1, 68n4
   mentorship of Rafael Lorente de Nó, 85–86
   photograph, 50f

structure of mammalian retina, 162f
   study of glial cells, 59–60
Cambridge Declaration on Consciousness, 47
Carlsson, Arvid
   photograph, 117f
   research on neurotransmitters, 116–18, 232
Carpenter, William
   debate on site of consciousness, 14–15
Caton, Richard, xiv–xv
   biography, 124b
   development of neuroscience, 2
   electricity and slow waves in the
      brain, 123–26
   and first electroencephalogram, 125
   and nervous-system electrical
      impulses, 13–14
   photograph, 124f
   recording of brain activity, 65
   visual cortex, mapping of, 79
cats, consciousness and cognitive
   ability in, 40
Cauller, L. J., somatosensory cortex, 210–11
centrencephalic debate, on brain sites and
   consciousness, 14–21, 15f, 29–30
cerebellum, cell types found in, 52f
cerebral cortex, 80f, 214f
   consciousness, relationship of motor and
      sensory areas to, 90–91
   cortical localization, first evidence of, 70
   and debate on site of consciousness,
      17–21, 29–30
   "grandmother cells," 29, 143–48
   hearing, taste, and smell, location of, 90
   Human Brain Connectome Project, 94
   isolation of, 103–5, 105f
   long-term memory, 92–93
   motor cortex, mapping of, 71–77
   origin of brain rhythms in, 134
   plasticity and, 94–98
   role in consciousness, 69
   role in creating sensation, 209–10
   sensory receptors, and specifying cortical
      modality, 236–37
   short-term memory, mapping of, 91–92
   single cell studies in, 134–36
   somatosensory cortex, mapping of, 82–90
   structure of, 110f
   thalamus, interactions with, 136–38, 220–23
   variability of cortical contributions to
      consciousness, 242–43
   visual cortex, mapping of, 78–82

cerebral cortex, wakefulness of, 103
   and function of brainstem, 103–5
   and hypothalamus, 118–20
   isolated cortex *(cerveau isolé)*, 103–5, 105*f*
   and locus coeruleus, 117–18
   main elements in control of wakefulness and sleep, 119*f*
   and neurotransmitters, 115–18
   nonspecific thalamic nuclei, 109*f*
   and nonspecific thalamic nuclei, 105–10
   reticular activating system, 110–15
Cerebri Anatome (Willis), 12, 13*f*
cerveau isolé (isolated cortex), 103–5, 105*f*
Chalmers, David
   consciousness in various forms of life, 254–56
   Montreal Symposium, 1997, 27–29
Charles Bonnet syndrome, 224–25n23
"Chaser," border collie, 40
children
   emergence of consciousness in, 4–5
chimpanzees, language in, 41–42
Chinese Room argument, and nonhuman consciousness, 261
Churchland, Patricia, qualitative aspects of sensation, 228
Clifford, W. K.
   on origins of consciousness, 35–36
Clynes, Manfred, CAT device, 188–89
cockatoo, consciousness and cognitive ability in, 39
Cohen, Stanley, and discovery of nerve growth factor, 54
Cold War, impact on neuroscience research, 21
"complex cells," among visual neurons, 168–69
computer of average transients (CAT), 188–89
computers, possibility of consciousness for, 254–63
concept cells
   among gnostic units, 203–4
   in hippocampus, 148–50
   See also "grandmother cells"
conditioning, and reflexive nature of human behavior, 229–36
conscious mental field (CMF) theory, 226–27
consciousness
   among variety of organisms, 1
   availability of neurons for, 44–47
   and cerebral cortex, wakefulness of, 103
   cortical columns and, 210–14
   defining, 4–5, 254
   double consciousness, 235–36
   and dualism *vs.* monism debate, 226–27
   early history of writings on, 12–14
   emergence in childhood, 4–5
   exhibited by various forms of life, 228, 254–57
   individual brain regions and contributions to, 242–45
   key observations on, 264–65
   key observations on, evidence for, 265–66, 266*t*
   major findings on, 200–1
   mind-body problem and, 5–11
   and modification of synapses, 227
   and patterns of brain activity, 138
   possibility for computers, 254–63
   relationship of motor and sensory areas to, 90–91
   sensory consciousness, evolution of, 241–42
   sensory systems and, 161
   sites contributing to, 14–21
   variability of cortical contributions to, 242–43
Consciousness, Confessions of a Romantic Reductionist (Koch), 159n12
consciousness, in animals
   availability of neurons for, 44–47, 45*f*, 46*t*
   Cambridge Declaration on Consciousness, 47
   Charles Darwin and *The Descent of Man,* 32–36
   crows, 39, 40*f*
   dogs and cats, 40
   dolphins, 42–43
   historical attitudes toward, 32, 47n1
   insects, 43*f*, 43–44
   neuroanatomy and neurophysiology, 45–47, 46*t*
   nonhuman primates, 41–42
   octopi, 40–41
   parrots, 38–39
   T. H. Huxley on, 36
consciousness, origins of
   W. K. Clifford on, 35–36

Coombs, Jack, slow waves as synaptic potentials, 132–33
corpus callosum, results of sectioning, 246–51, 249f
cortical columns
  activity within, 208f, 210
  and consciousness, 210–14
  critical excitation concept, 212–13
  and gnostic units, 205–10
  as gnostic units, 205–6
  number of gnostic units in, 213–14, 214f
  in somatosensory cortex, 208f, 210–12
  See also gnostic units
cortical impulse activity, and neurological disorders, 212–13
Courtois, G., origin of brain rhythms, 134
creativity, unexpected nature of, 235–36
Crick, Francis
  and development of neuroscience, 1
  and history of neuroscience, 25–26
  photograph, 26f
critical excitation concept, cortical columns and, 212–13
crows, consciousness and cognitive ability in, 39, 40f

Dale, Henry, on neurotransmitters, 115–16
Damasio, Antonio
  brainstem as site of consciousness, 29–30
  photograph, 29f
  primordial feelings of protoself, 244–45
Dandy, Walter, pioneering neurosurgery, 246
Darwin, Charles
  biography, 33b
  on cognitive ability in baboons, 41
  on cognitive ability in parrots, 39
  The Descent of Man and consciousness in animals, 32–36
  double consciousness, 235–36
  portrait, 33f
Da Vinci, Leonardo
  external features of the brain, 69
Davis, Hallowell
  and first EEG recordings, 106
  life and research, 167
Dawson, George
  biography, 187b
  life and research, 186–88
  photograph, 187f
deafness, studies of recovery from, 178

declarative memory, 92
Deecke, Lüder, 189–90, 191f
Dennett, Daniel, 69
Descartes, René
  biography, 7b
  coordination of muscle and visual mechanisms, 6f
  mind-body problem, 5
  portrait, 7f
  sense of self, origin of, 226–27
  Treatise on Man, 12, 123
Descent of Man, The (Darwin)
  on cognitive ability in parrots, 39
  and consciousness in animals, 32–36
Diamond, Jack, photograph, 95f
dogs, consciousness and cognitive ability in, 40
dolphins, consciousness and cognitive ability in, 42–43
Dostrovsky, Jonathan, "place cells" in rat hippocampus, 154–59
dualism *vs.* monism, 226–27, 228–29
  John Eccles, 8–10
  René Descartes and, 5
  Thomas Huxley and, 6–10
Du Bois-Reymond, Emil
  demonstration of nerve impulses, 60
  work on nerve impulses, 123

Eccles, John
  biography, 18–19b
  contributions of, 267
  dualism *vs.* monism, 8–10
  excitation and inhibition, studies of, 66
  as graduate student, xiv–xv
  on neurotransmitters, 115–16
  photograph, 23f
  portrait, 22f
  sense of self, origin of, 226–27
  Vatican Symposium, 1964, 24
  work on slow waves as synaptic potentials, 131–34, 139n13
Efron, Ron, backward masking experiments, 179
electricity
  and Luigi Galvani's experiments on brain, 12–13
  and Richard Caton's experiments on brain, 13–14

276 INDEX

electricity, and the brain
   and electroencephalogram (EEG), demonstration of, 131
   and electroencephalogram (EEG), development of, 126–28
   and electroencephalogram (EEG), recordings of, 130f
   A History of the Electrical Activity of the Brain, 139n4
   impulses *vs.* slow waves, 123–26
   interactions between thalamus and cortex, 136–38
   intracellular recordings from motoneurons, 131–33, 133f, 139n12
   readiness potential (Bereitschaftspotential), 189–92, 195–98
   signal averaging, development of, 186–89
   single cell studies in cortex, 134–36
   slow waves as synaptic potentials, 131–34
   work of Edgar Adrian at Cambridge laboratories, 128–31
electroencephalogram (EEG)
   first recordings, 106, 125
   history of, 127–28, 139n4
   recordings of, 130f
   work of Hans Berger on, 126–28
electrophysiologists, and mapping of visual cortex, 79–82
"encephalitis lethargica," treatment with L-dopa, 122n27
Enigma Machine, analogy of, 237–40, 238f, 239f
epilepsy
   electrophysiological recordings in patients, 148–50
   research by John Hughlings Jackson on, 74b
epiphenomenalism, arguments for and against, 10
episodic memories, 92
Evarts, Edward, intracellular recordings, 135
evolution, of sensory consciousness, 241–42
"exit units," and integrative activity of the brain, 152–53
explicit memory, 92
Eye, Brain and Vision (Hubel), 163
eyes, consciousness and visual system, 161–63

Faraday, Michael, 69
Feinstein, Bertram, 192

Ferrier, David, xiv–xv
   demonstration of brain activity, 65
   motor cortex, mapping of, 72–76
   portrait, 73f
   visual cortex, mapping of, 78, 82
fetal cortex, development of, 94–97
Flourens, Jean-Pierre
   motor cortex, mapping of, 71–72
Foerster, Otfried
   motor cortex, mapping of, 76–77
   neurosurgical innovations, 128
Forbes, Alexander
   and first neurobiology department, 167
   life and research, 106–7
   photograph, 106f
free will
   and synthesis of brain activity, 228–29
free will experiment, of Benjamin Libet, 195f, 196–97, 228–29
Freiburg, Germany, history of, 189
Fritsch, Gustav
   motor cortex, mapping of, 71–72
   portrait, 71f
Fulton, John
   Physiology of the Nervous System, ix, 82–84, 110–12
functional brain imaging
   and history of neuroscience, 25–26
functional magnetic resonance imaging (fMRI), 25–26

Gage, Phineas, role in history of neuroscience, 69–70
Galambos, Robert, 37
Galen
   external features of the brain, 69
Galvani, Luigi
   discovery of "animal electricity," 12–13, 60
   pioneering work on electricity and the brain, 123
Galvani's Spark (McComas), xi
galvanometer, and history of neuroscience, 13–14
ganglion cells, Stephen Kuffler study of, 163–66, 165f
Gardner, Allen and Beatrice, and chimpanzee, "Washoe," 41–42
Gasser, Herbert, work on slow waves, 123–25
Gazzaniga, Michael
   biography, 247b

photograph, 247f
sectioning of corpus callosum, 246–51, 249f
on sense of self, 251
Gerard, Ralph
  biography, 74–75b
  memory, from short-term to long-term, 91–92
  visual cortex, mapping of, 79–82
glial cells, 59–60
  average number of, 66–67
gnostic fields, anatomical locations o, 205
gnostic units
  anatomical locations of gnostic fields, 205
  for categories, 201–3
  concept cells, 203–4
  cortical columns and, 205–10
  cortical columns as, 205–6
  creation of new, 206–9, 207f, 208f
  and facial recognition, 203
  history and description, 201
  impulse activity, role of thalamus in, 220–23
  for individuals and objects, 203–4
  and integrative activity of the brain, 152–54
  interaction among, 215–16
  kinesthetic gnostic units, 204–5, 218–20
  numbers of, 213–14, 214f
  processing errors, 215
  reconsideration of function, 181–83
  and sense of self, 216–17
  time needed for creation, 215
  types of, 201–5
  and visual categories, 202f
  voluntary movements and, 218–20
  See also concept cells; cortical columns; grandmother cells
Golgi, Camillo
  neuroscience and staining methods of, 50–53
  photograph, 51f
Graham, Helen, work on slow waves, 123–25
"grandmother cells"
  in the human hippocampus, 148–50
  The Integrative Activity of the Brain (Konorski), 150–54
  introduction of concept, 29, 143–48
  and necessity for single neuron recordings, 141–43
  and "place cells" in rat hippocampus, 154–59
  See also concept cells
Granit, Ragnar
  as graduate student, xiv–xv
Grass, Albert, neuroscience equipment, 106
Gray, Charles, thalamocortical relationship, 222
Gray, John, ix
Grey Walter, William
  artificial nervous systems, design and building of, 257–59
  biography, 258b
  photograph, 258f
Grey Walter, William, readiness potential, 190–92
Griffin, Donald
  biography, 37b
  problem-solving in animals, 37–38
Gross, Charles, work on "grandmother cells," 147–48

Hamburger, Victor, 53–54, 68n7
Harvard University, history of medical education, 106
Head, Henry
  biography, 16b
  epicritic and protopathic nervous systems, 244–45
  parietal lobe and body schema, 87
  portrait, 13f
  thalamus, role in consciousness, 15–17
  with William Rivers, 16f
hearing, cortical location of, 90
Hebb, Donald
  life and research of, 58–59
  sensory deprivation experiments, 243
Helmholtz, Hermann
  demonstration of nerve impulses, 60
Herculano-Houzel, Suzana
  estimate of neurons in the brain, 66–67
Hillarp, Nils, research on neurotransmitters, 116–18
Hinton, Geoffrey
  artificial intelligence and "deep learning," 3
  life and research of, 263n4
  photograph, 260f
hippocampus
  concept cells in, 148–50
  "place cells" in rat hippocampus, 154–59

Hippocrates
  writings on the brain, 12
History of the Electrical Activity of the Brain, A, (Brazier), 139n4
Hitzig, Eduard
  motor cortex, mapping of, 71–72
  portrait, 71*f*
Hodgkin, Alan
  demonstration of nerve impulses, 61–62
  photograph, 63*f*
Hoffmann, Paul, investigation of spinal cord function, 128
honeybees, problem-solving among, 38, 43
Hooke, Robert
  Micrographia, 2–3
Hubel, David
  cortical columns, 205–6
  design of electrodes, 141–42
  Eye, Brain and Vision, 163
  with Herbert Jasper, 27*f*
  photograph, 166*f*
  somatosensory cortex, columnar organization of neurons in, 82–84
  vision studies, 2
  visual cortex, exploration of, 166–69
  visual cortex, parallel processing in, 169–70
  visual processing, alternative schemes, 171–72
human behavior, reflexive nature of, 229–36
Human Brain Connectome Project, 94
Huxley, Andrew
  demonstration of nerve impulses, 61–62
  photograph, 63*f*
Huxley, Thomas Henry
  biography, 7–8*b*
  consciousness, origins of, 227
  on consciousness in animals, 36
  dualism *vs.* monism, 6–10
  portrait, 9*f*
  response to Charles Darwin, 33–35
hypothalamus, and wakefulness of the cortex, 118–20

illustrations, availability of, xi
implicit memory, 92
insects, consciousness and cognitive ability in, 43*f*, 43–44
instrumental conditioning, descriptions of, 232
Integrative Activity of the Brain, The (Konorski), 3, 150–54

Jackson, Frank, nonhuman consciousness, 262
Jackson, John Hughlings
  biography, 74*b*
  motor cortex, mapping of, 72–75
  portrait, 74*f*
James, William
  convergence in sensory pathway, 145
  photograph, 219*f*
Jasper, Herbert, x
  biography, 23*b*
  collaboration with Wilder Penfield, 18*b*
  with David Hubel, 27*f*
  Laurentian Symposium, 1953, 21
  Montreal Symposium, 1997, 27–29
  nonspecific thalamic nuclei and cortex, 108–10
  portrait, 26*f*
  research meeting hosted by, 1
  Vatican Symposium, 1964, 24–25
  "JD," octopus at Zoological Station, Naples, Italy, 41
Jung, Richard, 189

Kandel, Eric, 59, 68n12
  molecular activity and creation of memory, 92
Kanizsa triangle, 177*f*, 177–78
Katz, Bernard
  Nerve, Muscle and Synapse, ix
kinesthetic gnostic units, 204–5
  voluntary movements and, 218–20
Koch, Christof
  life and research, 159n12
  photograph, 148*f*
  work with epilepsy patients and concept cells, 148–50
Kolliker, Albrecht, 51–52
Konorski, Jerzy
  biography, 151*b*
  on creation of new gnostic units, 209
  gnostic units, introduction of, 2
  gnostic units, research on, 201–5
  The Integrative Activity of the Brain, 3, 150–54
  meeting with, x–xi
  reflexive nature of human behavior, 231–32
  sensory "analyzer," 153*f*

Kornhuber, Hans
  life and research, 189–94
  photograph, 190f
  readiness potential, development of, 189–90, 191f
Krause, Feodor
  motor cortex, mapping of, 76–77
Kristiansen, K., origin of brain rhythms, 134
Krnjević, Krešimir
  life and research, 121n24
  neurotransmitters, 116
Kuffler, Stephen
  biography, 164b
  photograph, 164f
  study of ganglion cells, 163–66, 165f
Kulics, A. T., somatosensory cortex, 210–11
Kurzweil, Ray, artificial intelligence and "deep learning," 3, 262
kymograph, use in studies of brain activity, 65–66

language
  among dolphins, 42–43
  in chimpanzees, 41–42
Larionov, Vladimir
  cortical location for hearing, 90
Lashley, Karl, 54–57, 58–59
  cortical areas and contributions to consciousness, 242
Laurentian Symposium, and history of neuroscience, 21, 22f
L-dopa
  and "encephalitis lethargica," 122n27
  treatment for parkinsonism, 118
learning
  artificial intelligence. ability of, 257–61
  by unconscious inferences, 174–76
Lettvin, Jerome
  biography, 144b
  and example of octopus intelligence, 41
  Lettvin's fable, 146b
  with Walter Pitts at MIT, 144f
Lettvin, Jerry
  and "grandmother cells," 143–45, 147
Levi-Montalcini, Rita
  biography, 55b
  and discovery of nerve growth factor, 54
Leyton, Albert, xiv–xv
  motor cortex, mapping of, 76–77
  primate motor cortex, 97

Libet, Benjamin
  biography, 193b
  brain activity and conscious awareness, 2
  and concept of free will, 228–29
  life and research, 192–95
  photograph, 193f
  pioneering experiments of, 192–98
  sense of self, origin of, 226–27
  Vatican Symposium, 1964, 25
light, and visual processing in cortex, 172
linear stimuli, visual neurons and, 167–68
Livingstone, Margaret, visual processing in cortex, 171–72
Llinás, Rodolfo
  biography, 221b
  portrait, 221f
  thalamocortical relationship, 136, 220, 222–23
locus coeruleus, and wakefulness of cortex, 117–18, 121–22n26
Lømo, Terje
  creation of memory, 92
Lorente de Nó, Rafael
  biography, 83b
  cortical columns, 205–6
  portrait, 85f
  somatosensory cortex, columnar organization of neurons in, 82–84
Lorenz, Konrad
  consciousness in animals, 38
Lucas, Keith
  demonstration of nerve impulses, 60–61
Ludwig, Carl
  invention of kymograph, 65–66

Mackinnon, Roderick, xi
  demonstration of nerve impulses, 62–63
Magoun, Horace
  biography, 111b
  discovery of reticular activating system, 110–15
Man on His Nature (Sherrington), 147
mapping of the brain
  cerebral cortex, role in consciousness, 69
  consciousness, relationship of motor and sensory areas to, 90–91
  cortical localization, 70
  external features of the brain, 69
  hearing, taste, and smell, location of, 90
  Human Brain Connectome Project, 94

mapping of the brain (*cont.*)
    long-term memory, 92–93
    motor cortex, 71–77
    plasticity and, 94–98, 97*f*
    short-term memory, 91–92
    somatosensory cortex, 82–90
    visual cortex, 78–82
Matthews, B. H. C.
    plasticity and mapping of the brain, 94
Matthews, Bryan, invention of
    oscillographs, 130–31
McComas, Alan J.
    Galvani's Spark, xi
McGinn, Colin, qualitative aspects of
    sensation, 228
McIlwain, Hugh, single cell studies in
    cortex, 135
Mechnikov, Ilya
    work on immunology, 63
memory
    intensity of experience and formation of, 209
    long-term memory, 92–93
    short-term memory, 91–92
Micrographia (Hooke), 2–3
migraine headache
    cortical impulse activity and, 212–13
    and distortions of the body schema, 88, 89*f*
    loss of body schema in, 243–44
    and visual processing in cortex, 171
Milner, Peter, on concept cells, 203
mind-body problem
    John Eccles and, 8–10
    René Descartes and, 5
    Thomas Huxley and, 6–10
Molaison, Henry (Patient HM), and
    localization of memory, 93
molecular activity, and creation of
    memory, 92
Molnár, Zoltán
    fetal cortex, development of, 94–97
monism *vs.* dualism, 226–27, 228–29
    John Eccles, 8–10
    René Descartes and, 5
    Thomas Huxley and, 6–10
Montreal Symposium, 27–29
Morison, Robert
    life and research, 120n11
    nonspecific thalamic nuclei and cortex, 107, 134

Moruzzi, Giuseppe
    biography, 113*b*
    discovery of reticular activating
        system, 112–15
    photograph, 113*f*
motor commands, and area 7 of
    somatosensory cortex, 89–90
motor cortex
    and mapping of the brain, 71–77
Mountcastle, Vernon
    biography, 81*b*
    cortical columns, 205–6
    kinesthetic gnostic units and analysis, 204
    photograph, 83*f*
    somatosensory cortex, columnar
        organization of neurons in, 82–84
    somatosensory cortex, mapping of, 89–90
movement
    ability to cancel, 197–98
    as cue to object recognition, 177–78
Munk, Hermann
    visual cortex, mapping of, 78
muscle and visual mechanisms
    coordination of, 6*f*

Nagel, Thomas
    consciousness exhibited by various forms
        of life, 228
    "What Is It Like to Be a Bat?", 44
nerve growth factor (NGF), discovery of, 54
nerve impulses, 60–63
nervous system
    discovery of electric impulses in, 12–14
neural net approach, to building artificial
    intelligence, 259–61
neuroanatomy
    human and animal, 45–47, 46*t*
    See also brain, mapping of; brain,
        neuroanatomy and neurophysiology of
neurological patients, impulse activity
    in, 212–13
neurons
    average number of, 66–67
    drawing by Cajal, 52*f*
    motoneurons, intracellular recordings
        from, 131–33, 133*f*, 139n12
    three types of, 53*f*, 135–36
neurons, availability for consciousness, 44–47, 45*f*, 46*t*
neurons, basic science of, 49–59

INDEX   281

discovery of neural pathways, 54–57
life and research of Santiago Ramón y
   Cajal, 49–53
life and research of Victor
   Hamburger, 53–54
nerve growth factor (NGF),
   discovery of, 54
synapse, discovery and introduction
   of, 57–58
synapse, usage and structure, 58–59
neurophysiology
   human and animal, 45–47, 46t
   personal study of, ix
   See also brain, mapping of; brain,
      neuroanatomy and neurophysiology of
neuroscience
   development of, 1–3
neuroscience, history of
   early writings on brain, 12–13
   Francis Crick, 25–26, 26f
   functional brain imaging, 25–26
   impact of Cold War on, 21
   Laurentian Symposium, 1953, 21, 22f
   Montreal Symposium, 1997, 27–29
   and nervous-system electrical
      impulses, 12–14
   Vatican Symposium, 1964, 24–25
neurotransmitters
   and wakefulness of cortex, 115–18
nonhuman primates, consciousness and
   cognitive ability in, 41–42

object recognition
   movement as cue to, 177–78
   visual processing and, 174
octopi
   consciousness and cognitive ability
      in, 40–41
   neuroanatomy and neurophysiology
      of, 46–47
O'Keefe, John
   biography, 155b
   photograph, 151f
   "place cells" in rat hippocampus,
      154–59, 160n18
"On the Electrical Excitability of the
   Cerebrum" (Fritsch & Hitzig), 72
*On the Hypothesis that Animals Are
   Automata, and Its History* (Huxley), 36
operant conditioning, descriptions of, 232

Organization of Behavior, The (Hebb), 58–59
Origin of Species, The (Darwin), 33–35
oscillographs, and research on electricity and
   the brain, 130–31

pain, ability to experience, 244–45
parallel processing, in visual cortex, 169–71
parkinsonism
   electrode use in patients, 142
parkinsonism, treatment with L-dopa, 118
parrots, consciousness and cognitive abilities
   in, 38–39
Patton, Harry, intracellular recordings, 134
Pavlov, Ivan
   biography, 230b
   portrait, 230f
   reflexive nature of human
      behavior, 230–32
   work on conditioned reflexes, 63
Penfield, Wilder
   biography, 18b
   and debate on site of consciousness, 18–20
   memory, localization of, 92–93
   motor cortex, mapping of, 76–77
   photograph, 18f
   portrait, 15f
   somatosensory cortex, mapping of, 82
   Vatican Symposium, 1964, 24–25
Pepperberg, Irene, and African Grey parrot,
   "Alex," 38–39
"Petra," African Grey parrot, 39
Phillips, Charles
   intracellular recordings, 134–35
   motor cortex, mapping of, 77
Physiological Society, United Kingdom
   attendance at meetings, ix
physiology
   research methodology, history of, ix–x
Physiology of the Nervous System (Fulton),
   ix, 82–84
   on reticular formation, 110–12
Pilley, John, and border collie, "Chaser," 40
Pitts, Walter, with Jerome Lettvin at MIT, 144f
"place cells," in rat hippocampus, 154–59
plasticity
   and mapping of the brain, 94–98, 97f
   neuroplasticity of fetal cortex, 236–37
pleasure, ability to experience, 244–45
"pontifical cells," and convergence in sensory
   pathway, 147

positron emission tomography (PET)
  and visualization of neurotransmitters, 25
Powell, Tom
  somatosensory cortex, columnar
    organization of neurons in, 82–84
precentral gyrus, and debate on site of
  consciousness, 18–20
primates, nonhuman, consciousness and
  cognitive ability in, 41–42
procedural memory, 92, 93
psychons, introduction of concept, 8–10
publications on consciousness
  increase in, x
Purpura, Dominick, cortex and
  thalamus, 136
Purves, Dale, visual processing and
  unconscious inference, 177–78

quadriplegia, brain activity in, 120n1
Question of Animal Awareness, The (Griffin), 38
Quiroga, Rodrigo
  cortical "grandmother cells," 29
  critical excitation concept, 212
  photograph, 150*f*
  work with epilepsy patients and concept
    cells, 148–50, 149*f*

radial glia, 59–60
rats
  consciousness of, 44
  "place cells" in hippocampus, 154–59
readiness potential *(Bereitschaftspotential)*,
    189–92, 191*f*
  commentary on, 195–98
  cortical processing time, 196
  free will experiment, 195*f*, 196–97, 228–29
  intended movement, ability to
    cancel, 197–98
  simplicity of tasks studied, 198
  supplementary motor area (SMA), 197
  visual processing time, 197
"Reflexes of the Human Brain"
  (Sechenov), 230–31
research
  originality of, xi
  perspectives on history of, 67
research methodology
  history of, ix–x
reticular activating system, discovery
  of, 110–15

retina
  anatomical studies of, 161–62
  ganglion cells, study of, 163–66, 165*f*
  structure of, 162*f*
  types of cells in, 163
Rivers, William
  collaboration with Henry Head, 16*b*
  with Henry Head, 16*f*
  portrait, 13*f*
  thalamus, role in consciousness, 15–17
Rolls, Edmund
  gnostic units and facial recognition, 203
  "grandmother cells," 147–48
  object recognition, 174

Sacks, Oliver, 122n27
Searle, John, and Chinese Room argument, 261
Sears, Thomas, thalamic nuclei, 134
Sechenov, Ivan
  life and research of, 64–65
  portrait, 64*f*
  reflexive nature of human behavior, 229–32
semantic memories, 92
sensation, qualitative aspects of, 228
sense of self
  and differences between brain
    hemispheres, 251
  and dualism *vs.* monism, 226–27
sense of self, gnostic units and, 216–17
sensory consciousness, evolution of, 241–42
sensory deprivation, effects of, 243
sensory receptors, and specifying cortical
  modality, 236–37
sensory substitution, and mapping of
  cerebral cortex, 97–98
sensory systems
  consciousness and, 161
Sherrington, Charles Scott
  biography of, xiv–xv
  on brain activity, synthesis of, 226
  contributions of, 267
  excitation and inhibition, studies of, 65–66
  experience of consciousness in animals, 41
  lifetime achievements, xiv
  motor cortex, mapping of, 76–77
  on nerve impulses, 60
  photograph, 106*f*
  "pontifical cells," and convergence in
    sensory pathway, 147
  portrait, xv*f*

INDEX 283

primate motor cortex, 97
synapse, introduction of term, 57–58
synapse, studies of, 63
vignette from life of, xiii–xiv
signal averaging, development of, 186–89
"simple cells," among visual neurons, 168–69
Singer, Wolf, thalamocortical relationship, 222
single neuron recordings
    necessity for, 141
Skinner, B. F.
    biography, 233*b*
    at Harvard, 233*f*
    reflexive nature of human behavior, 232
    Skinner Box, 232, 235*f*
skin sensation, mediation of, 16*f*, 16–17
sleep, main elements in control of, 119*f*
smell, cortical location of, 90
"Snowball," cockatoo, 39
somatosensory cortex
    cortical columns, investigations of, 208*f*, 210–12
somatosensory cortex, mapping of, 82–90, 88*f*
    area 7 and motor commands, 89–90
    body schema, 87
    body schema, clinical distortions of, 88
    columnar organization of neurons in S1, 82–84
    higher somatosensory processing, 86–87
Spemann, Hans, 53–54
Sperry, Roger
    biography, 56*b*
    life and research of, 54–57, 57*f*
    research on split brain, 246
    sectioning of corpus callosum, 246–51
    Vatican Symposium, 1964, 25
Spinoza, Baruch, and concept of free will, 228–29
Steriade, Mircea, 2
    biography, 129*b*
    photograph, 137*f*
    thalamocortical relationship, 136–38, 220
Studies in Neurology (Head), 15–17
supplementary motor area (SMA), 197
synapse
    components of, 58*f*
    discovery and introduction of, 57–58
    excitation and inhibition at the, 63–66
    modification of synapses, 227
    usage and structure of, 58–59
synaptic changes, and creation of memory, 92

synaptic potentials, slow waves as, 131–34
synchronous impulse firing, and parallel processing in visual cortex, 170–71

taste, cortical location of, 90
thalamus
    cerebral cortex, interactions with, 136–38
    and debate on site of consciousness, 14–17, 15*f*
    electrode recording of activity in, 142*f*, 142–43
    nonspecific thalamic nuclei, 105–10, 109*f*
    and origin of brain rhythms in cortex, 134
    role in impulsive activity of gnostic units, 220–23
"time chunks," visual processing and, 179–81, 182*f*
Tinbergen, Nikolaas
    consciousness in animals, 38
"transit units," and integrative activity of the brain, 152–53
Treatise on Man (Descartes), 12, 123
T'sao, Doris, work on object recognition, 174
Turing, Alan, 254, 263n1
    biography, 255*b*
    photograph, 255*f*

unconscious inferences, learning by, 174–76
"unitary events," and integrative activity of the brain, 153–54

Van Essen, David, Human Brain Connectome Project, 94
Vatican Symposium, and history of neuroscience, 24–25
Vesalius, external features of the brain, 69
Virchow, Rudolf, xiv–xv
    discovery of "neuroglia," 59
visual and muscle mechanisms
    coordination of, 6*f*
visual cortex
    enlarged role of, 178–79
    Hubel and Wiesel's exploration of, 166–69
    mapping of, 78–82
    parallel processing in, 169–71
    rhythms generated in, 130–31
    slow wave responses in, 131
    visual processing, alternative schemes, 171–72, 174*f*
visual processing time, 197

visual neurons
  hierarchy of, 168f
  linear stimuli and, 167–68, 169f
visual system, consciousness and, 161, 183
  backward masking, 179–81, 182f
  blindness, studies of recovery from, 177–78
  the eye, 161–63, 162f
  gnostic units, function of, 181–83
  learning by unconscious inferences, 174–76
  linear stimuli and visual neurons, 167–68
  object recognition, 174
  parallel processing in visual cortex, 169–71
  processing in visual cortex, alternative schemes, 171–72, 174f
  retina, structure of, 162f
  Stephen Kuffler's study of ganglion cells, 163–66, 165f
  "time chunks," and visual processing, 179–81, 182f
  visual cortex, enlarged role of, 178–79
  visual cortex, Hubel and Wiesel's exploration of, 166–69
  visual stimuli, 168–69
Vogt, Oskar and Cecile, neurological institute created by, 78–79, 128
voluntary movements, gnostic units and, 218–20
Von der Malsburg, Christof, thalamocortical relationship, 222
Von Frisch, Karl
  consciousness in animals, 38
Von Helmholtz, Hermann
  biography, 175b
  learning by unconscious inferences, 174–76
  photograph, 175f
Voyage of the Beagle, The (Darwin), 33–35

wakefulness, main elements in control of, 119f
Waking Brain, The (Magoun), 111b
Wall, Patrick
  biography, 157b
  photograph, 155f
  "place cells" in rat hippocampus, 154–59
Walshe, Francis, x
  and debate on site of consciousness, 20
Walton, John, ix
"Washoe," chimpanzee and student of American Sign Language, 41–42
Watson, James, 25
Watson, John
  life and work of, 252n10
  reflexive nature of human behavior, 232
Wearing, Clive (Patient CW), and sparing of procedural memory, 93
Weiss, Paul, 54–57
"What Is It Like to Be a Bat?" (Nagel), 44
Wiesel, Torsten
  cortical columns, 205–6
  photograph, 166f
  somatosensory cortex, columnar organization of neurons in, 82–84
  vision studies, 2
  visual cortex, exploration of, 166–69
  visual cortex, parallel processing in, 169–70
Willis, Thomas
  on cerebral hemispheres, 69
  Cerebri Anatome, 12, 13f
  external features of the brain, 69
Woodbury, Walter, intracellular recordings, 134
Woolsey, Clinton
  somatosensory cortex, mapping of, 82
Wren, Christopher, 12, 13f
  etching by, 13f

Zeki, Semir, parallel processing in visual cortex, 169–70
Zoological Station, Naples, Italy, octopus at, 41

www.ingramcontent.com/pod-product-compliance
Ingram Content Group UK Ltd.
Pitfield, Milton Keynes, MK11 3LW, UK
UKHW021250180426
11946UKWH00003B/62